THE *SMART* SMART HOME HANDBOOK

An Hachette UK Company
www.hachette.co.uk

First published in Great Britain in 2018 by Ilex, an imprint of
Octopus Publishing Group Ltd
Carmelite House
50 Victoria Embankment
London EC4Y 0DZ
www.octopusbooks.co.uk
www.octopusbooksusa.com

Distributed in the US by Hachette Book Group
1290 Avenue of the Americas, 4th and 5th Floors, New York, NY 10104

Distributed in Canada by Canadian Manda Group
664 Annette Street, Toronto, Ontario, Canada M6S 2C8

Publisher, Photo and Tech: Adam Juniper
Editorial Director: Helen Rochester
Managing Editor: Frank Gallaugher
Senior Editor: Rachel Silverlight
Publishing Assistant: Stephanie Hetherington
Art Director: Julie Weir
Designer: Cassia Friello
Cover Design: Eoghan O' Brien
Senior Production Manager: Peter Hunt

ISBN 978-1-78157-580-2

A CIP catalogue record for this book is available
from the British Library.

Printed and bound in China

10 9 8 7 6 5 4 3 2 1

THE *SMART* SMART HOME HANDBOOK

CONNECT, CONTROL & SECURE YOUR SETUP THE EASY WAY

ADAM JUNIPER

ilex

CONTENTS

BASICS

ECOSYSTEMS

SMART HOME SYSTEMS

ENTERTAINMENT

TEMPERATURE

LIGHTING

KITCHEN

MONITORING & SECURITY

WELL-BEING

BOTS & SEQUENCING

RELIABLE SETUP

SECURITY & PROBLEM-SOLVING

INTRODUCTION

Smart home technology enables a world in which that nebulous tech-industry expression "the Internet of Things" (also known as IoT) suddenly refers, very specifically, to *our* things. And not just our laptops and our phones, but our lights, our light switches, our thermostats, security systems, power sockets, doorbells, and vacuum cleaners. To name just a few . . .

The idea of making our homes "smart" is far from new, but in the past it has only been achieved at great expense and not always entirely elegantly, let alone in a way that the majority of us can replicate. The famous British racing driver Stirling Moss was a great enthusiast of home automation, but the bespoke arrangement of motorized pulleys and trays that made his breakfast appear from the ceiling still depended on there being someone at the other end to put the meal on the tray in the first place.

That's not to say that there haven't been some notable successes along the way. The Monsanto House of the Future, part of Disneyland California's Tomorrowland experience in 1957, gets a big "bing!" for predicting the arrival of the microwave in our homes.

The more recent past has also seen a few fairly confident steps toward home automation — not least with The Clapper light switch, which started listening for commands back in 1986, comfortably beating Amazon's Alexa by almost 30 years.

Technology is much more exciting now; it is a lot more advanced, as well as generally being cheaper and easier to bring into the home without any hassle. Although, if you are renovating a home or own the kind of new build that we see with a mix of enthusiasm and jealousy on TV shows like *Grand Designs* and *Cribs*, there are still built-in systems such as the KNX Bus or Control4 that require the fitting of dedicated cabling.

This book focuses on the sort of systems that we can buy off the shelf, those that are realistic for the average home and don't need to be installed all in one go. This is the area of the market that has seen the most growth in the last few years, and that is not expected to stop. But this is not the refined world of whole-home systems; it is one in which we mere humans are required to make regular choices, with uncertain outcomes. Assaulted on all sides by promotions, we need to ask ourselves questions such as, "do we actually want to be able to switch on or off our lights and change their color remotely?" And even where we think that's cool, would we give up the traditional switch? We'll talk about these issues, which can be labeled the "beta problem," a bit more overleaf.

So that's what this book is all about: helping you build the smart home of your dreams by showing you what's possible, making recommendations, then anticipating any problems and heading them off at the pass.

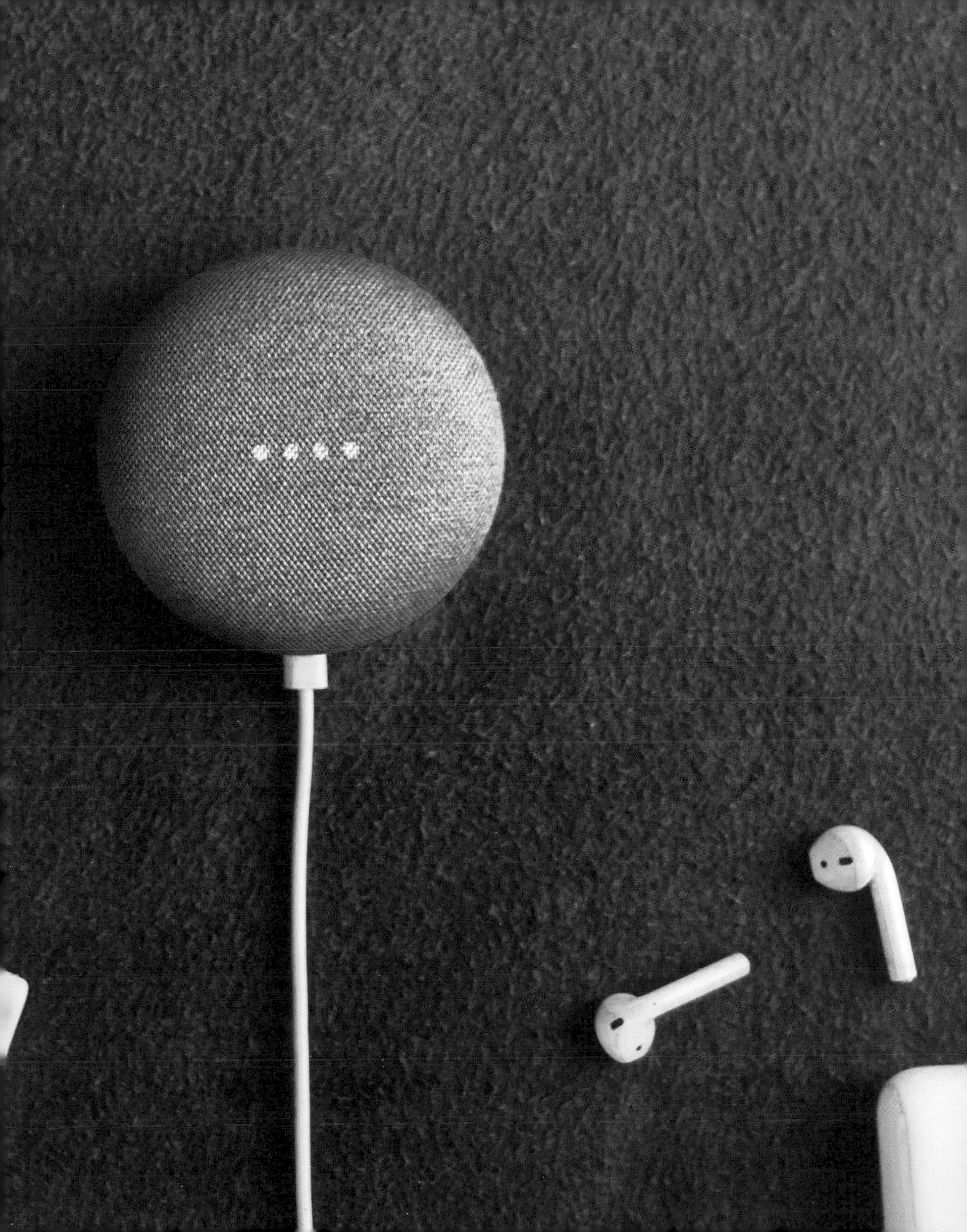

THE BETA PROBLEM

Not everyone wants to be a tester . . .

If you're interested in having a smart home, it's likely you're familiar with the name "beta tester" — the tech industry's term for someone using an experimental version of their software or hardware before it goes on sale to help identify any bugs or glitches.

Traditionally, beta testers report back to the developer in return for some kind of incentive, though often people do it just for the kudos of being the first to use a new product. From a certain point of view, being a beta tester is quite exciting, as some companies require the tester to sign a non-disclosure agreement, which leaves said tester with a tingling sensation of power and importance. Once the developer has received the feedback, they use that information to correct any issues that have been pointed out and thus edge the finished product toward perfection.

Back in 2004, Google's Gmail was launched into a beta limbo that lasted about five years, during which time it became one of the world's most popular email providers. Now we're almost used to Silicon Valley's offerings being in a kind of permanent beta, with Facebook or Instagram adding new features all the time, often with very little fanfare.

In the home, though, we're more familiar with the things we have around us that have worked perfectly well for decades. The light switch has not seriously changed in a century (The Clapper aside, but who actually has one of those these days?), and the upshot is that its function is a natural reflex for people, and tampering with such an intuitive operation should come with a risk warning attached.

I'm going to make the assumption that if you're reading this book you're at least interested in the possibilities of home automation, if not actually excited at the prospect of having your lights operated by voice command. Adopting this technology can bring some of the adventure of being a beta tester, but ask yourself, does that same thrill extend to everyone in your home?

The technology that makes the smart home work is often called the Internet of Things, and make no mistake, even though you can't see screens and keyboards on the items in question, they do contain software, require network connections, and, like your phone or your computer, will occasionally need software updates.

As you might imagine, your humble author is, of course, quite enthusiastic about this (and many other kinds of) technology. So in order to make this book truly helpful I've asked a few other people living in newly smartened homes about their experiences — whether they were the victims of an enthusiastic installer, or were a tech fan themselves. Over the next few chapters you can meet them, learn their thoughts and their passion for the smart home, and maybe even take some of their ideas with you.

I just want everything to work; with a toddler running around the house I don't have time to troubleshoot gadgets. I certainly don't want to be paying for them!

As a father with school-age children, I enjoy fun as much as they do, but I'm starting to worry about how tech affects them.

I live on my own and I do love gadgets; I've had some really great ones. I guess that means that I'm pretty picky though — I don't like it when machines behave badly.

I don't mind technology helping out, and I've got plenty of time to fix it when things go wrong.

WHAT MAKES TECH "SMART?"

We've lived with technology for a long time. What's changed?

In the '80s, we welcomed computers into our homes, but they didn't get a say in what temperature the thermostat was set to. Since then, computing has been on a journey; one that took in connectivity through the internet, then portability, and eventually pocketability in our phones. With all the miniaturization that entailed, it soon became easy to add internet connectivity to any digital device.

Digital technology is not new to the home; we're used to electronic timers on our ovens and porch lights activated by sensors. All of this might have been considered a little smart in the '80s, but even then the more ambitious consumers were looking at something called an X10 system (a technology developed in the late '70s that used the home's own wiring, much like some internet range extenders do today). These days it is generally accepted that a "smart home" device is one that extends control to an app on a smartphone or other device. That might be as simple as a remote on-off switch, but the relative ease of adding extra controls (like a timer) means that typically these app-enabled devices tend to be pretty feature-packed when compared to the "dumb" alternatives they are replacing.

It's worth noting that smart home functionality has typically come at a premium, so users have understandably expected devices to match the capabilities of previous top-of-the-range equivalents, as well as adding smartness. However, that is starting to change; the smart home concept has moved past the crucial early-adopter phase in which manufacturers could charge whatever the market would bear, and consumers are now seeking smart functionality without the premium price tag. An early example of this came from Nest, the company known above all for their heating thermostat, which is often cited in lists of the first smart home devices.

Smartness is also inferred from how the device, and its accompanying app, can respond without command. You could always simply turn your heating system on when you got home, or set a timer so that it's warm on your return. What makes the technology "smart" is when the heating system can detect you approaching home (via your phone's GPS) and turn the heating on for you, so it'll be nice and warm the moment you walk through the door, even if you're back early; and it won't use energy unnecessarily if you're late.

The next step to smartness in home devices is the ability to respond to natural-language commands on their own or as part of a group, for example, through Amazon's Alexa. It is devices with this kind of broad compatibility that are worth considering now.

It was a Hong Kong Shue Yan University study that identified the next step for smart devices: a robot buddy who interacts with these devices on your behalf, as well as taking photos that are automatically stored for you, and even calling the emergency services if needed. For those ready to invite a plastic pal into the home, the good news is that many devices available now will work with them.

NEST E VS NEST 3G

The Nest, at the time of writing on its third generation (above right), is one of the most iconic devices in the smart home world. The Nest E (above left) is the economy version, with reduced functionality and using cheaper materials to offer a lower barrier to entry.

OVEN TIMERS

Once upon a time, a digital timer that could switch off the oven was pretty smart.

ZENBO

Robots with voice recognition and artificial intelligence like Zenbo are the next generation in the smart home.

THE ESSENTIALS

Before you choose your smart home kit, let's get back to basics

Because, as I mentioned, smartness in devices is somewhat open to debate, the first thing you'll need to do is take a quick refresher on how devices are typically arranged, and the services they require, so you will be able to ensure smooth operation.

One of the most familiar functions of the smart home is the ability to activate assorted devices via voice assistants fitted in speakers and other equipment. It's important to look at these platforms, or "ecosystems," so we'll cover the various options in some detail. First we need to dispense with one other piece of jargon (one which is very commonly misused): the word "hub."

Hub is a nice, friendly word that suggests being at the center of things, so it's often associated with the speakers or digital assistants that get a lot of air time. In fact, a hub is not the same thing at all (except, irritatingly for the author, in the case of the Alexa Echo Plus, which we'll come to later), and it isn't central. A hub is an essential piece of equipment, but one which tends to be kept more or less out of sight.

Many of your devices will receive their commands via Wi-Fi from the router which also controls your home internet. Complications arise because other devices receive commands through different channels, i.e., not ones which your router is adapted to deal with. This is where the hub comes in: it speaks the language of those devices and acts as a translator between them and your router, so that you can control them from your phone or tablet (via an app).

So while it's tempting to see your digital assistant/smart speaker as the heart of the arrangement, giving and receiving commands directly, this is not the case at all. A typical arrangement of smart home technology is more like the one illustrated opposite, in which everything is relayed back to the router, sometimes via one or more additional hubs. When you ask Alexa or Siri to turn out the lights, for example, this command is relayed by Wi-Fi to the router, which speaks to the hub that controls the lights, and — if you got the command right — the lights go out. Even if everything operates on Wi-Fi and doesn't need a hub, the command still goes via the router.

You'll also hear about "interoperability." This refers to what you can (or cannot) do to use one system with another. That doesn't just mean, for example, can I ask Alexa to turn on this brand of lights, but also larger questions like, can I set this group of smart devices to work together?

This book covers a lot of approaches, and the benefits of Wi-Fi versus hub-based systems will come up again when we get into the specific tech, but for now it's important to get to grips with the basics, because they will be crucial to understanding what it is that you're actually acquiring.

SMART SPEAKER

The smart speaker listening for your commands is probably connected via Wi-Fi.

SMART DEVICES

Here are some smart bulbs that operate over a Z-Wave network (for example). They receive their command from the hub.

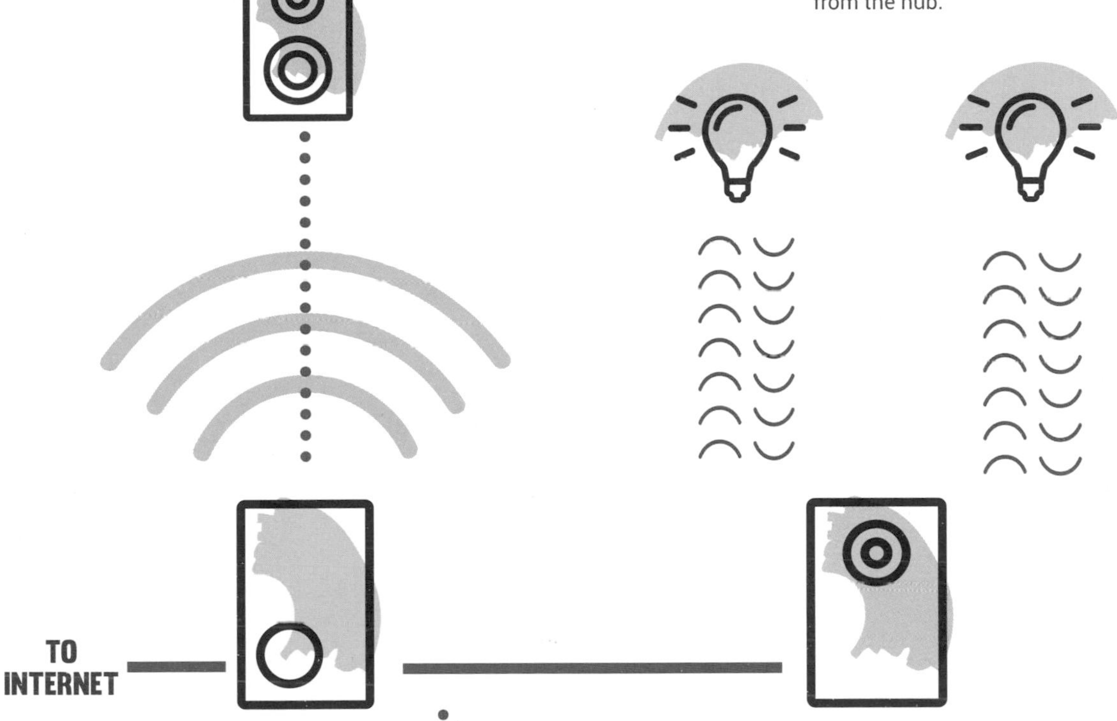

MODEM/ROUTER

Your connection to the internet, which is likely also your Wi-Fi router.

HUB

Device that relays commands to smart home devices based on a non-Wi-Fi communication standard.

TABLET/PHONE

Connected to your Wi-Fi it can be used to send commands to other devices, via the router.

SMART THERMOSTAT

Connected via Wi-Fi.

WHERE TO GO NEXT

For speed readers, or those impatient to get going!

I've attempted to provide those who are completely new to smart home tech with all the information they'll need to get started, while also offering a lot of useful advice to those who are a bit more tech-savvy. So read on and see where it's best for you to get started; if you know the basics already, you can skip ahead to the information that's most relevant to you without missing anything essential.

BEGINNING:
I'M SMART HOME CURIOUS BUT I DON'T KNOW A THING ABOUT IT

If you're starting from scratch and really don't know where to start, the beginning is as good as any place! Turn over the page and start learning about the major players in the smart home world. Or maybe you'd just like to know what kind of technology is available? In which case you might just want to skim through pages 52–129 to see what's currently out there.

GETTING SET UP:
I WANT TO LEARN ABOUT ALEXA AND OTHER SMART VOICE ASSISTANTS

So you're ready to go and take your home into the future – all you need to do now is pick a digital assistant, right? By all means turn over to Ecosystems to ponder Alexa and Siri's pros and cons. But before you dive in, it will also be worth checking you really understand How Your Internet Works (page 30), as well as the other networks that you might want to connect to it (pages 32–41).

146

SEQUENCING: I WANT MY HOME TO DO REALLY SMART THINGS

This is all about connecting your different devices and systems to make everyday living a breeze. So, for example, you can get out of bed and the coffee pot will already be boiling and the shower water will be hot.

TROUBLESHOOTING: SOLUTIONS FOR COMMON PROBLEMS

Let's face it, technology isn't always perfect. If something is not working you properly, either re-read the section on whatever device it is that is failing, or head here for more suggestions.

128

28

16

ECOSYSTEMS

This chapter introduces the major players, and the digital assistants that offer you voice control. You'll also gain some insight into the battle of the brands vying to help you control your home, and what each can offer you.

AMAZON ALEXA & ECHO

Perhaps the most logical company to offer a smart assistant

Retail giant Amazon's technology announcements can broadly be divided into two categories: those likely to pay off and those that make people say "awesome!" Amazon's digital assistant, Alexa, and the Echo smart speaker range are very much the former; perhaps the most innovative way a retailer has ever reached and understood its customers.

Of course, you need an Amazon account to use Alexa. So before you dive in, you should know that Amazon uses Alexa searches to gather data, and you can bet that, by using it, Amazon is going to learn a lot about you. The flip side of this is that Alexa's voice-recognition system has to be useful enough for you to invite "her" into your home in the first place.

(Most of us think of Alexa as "her," but the name is actually in homage to the great Library of Alexandria, and the activation word can be changed to other things, including "Computer" — which certainly feels right for those of us who grew up with *Star Trek: The Next Generation.*)

Alexa's usefulness comes not from its ability to order products for you without you needing to reach for a screen, but from the number of other tasks Amazon allows it to perform. Just as a smartphone comes with a number of apps built in — some for essential functions and some non-essential extras added to give the phone an edge over its competitors — any Alexa device is capable of performing some basic operations, for example, telling you today's weather or playing a podcast. Its functionality can be extended further by the addition of "Skills" — apps from third parties, from the patently useful, such as the Nest Thermostat (see page 58), to entertaining games — which Amazon has ensured are easy to develop (easy by software developers' standards).

The installation of these functions and Skills is achieved through your smartphone via an app linked to your Amazon account. Once set up, you can manage them either through your phone, or by giving the appropriate voice commands to your Echo.

Echo is Amazon's smart speaker brand, with products ranging from the highly affordable but perfectly functional Echo Dot through to the higher-end Echo Plus, which features

Alexa, what's the weather going to be like this week?

An Echo is constantly listening for the keyword (usually Alexa). When it hears it, it sends the next sentence to Amazon's computers for processing. There the system analyzes the sentence for keywords – here it would spot "weather" and "week." It will then respond with a list of predicted highs, lows and conditions for the coming seven days; leave off the word week and you'll just get the day's prediction, which your window may be just as able to serve.

Alexa, tell Harmony to turn on the TV.

With some Skills (optional features you have added), Amazon requires you to invoke them using the product's name, and this varies depending on region. Harmony, for example, is a Logitech product that can replace TV remotes. Here the voice-recognition system spots the word "tell" and "Harmony." It will then look for the "TV" and "On" elements within the Harmony Skill. The downside is that remembering the branding of Harmony is irksome and unnatural.

ECHO AND ECHO DOT

Amazon's Echo, of which the one shown (opposite) is 2017's second generation, was arguably the device that launched a thousand digital assistants. The Echo Dot (right), Amazon's first sub-$50 smart speaker, made Alexa a very easy purchase for many.

more powerful speakers and a built-in hub to optimize its compatibility. The product is capable of user switching by voice command, so different users can activate functions specific to them – for example, asking about their own daily schedule, or telling Alexa to play their music.

You don't need an Echo to use Alexa. It's possible to use Alexa from your computer or phone – even on an iPhone, via the Reverb app. In fact, you can connect with Amazon's assistant from any device that has it installed, such as your smart thermostat in the hallway. You can use nearly all the commands you would through an Echo, controlling blinds and lights with abandon; the only restrictions on these extended manifestations of Alexa is that they're barred from playing music from the Amazon Music service.

Alexa has also moved beyond the speaker into the world of the visual. New devices such as the Echo Show (see opposite) can present you with visual responses to what you would normally hear from Alexa.

The only real disappointment of Amazon's product range is Alexa's reluctance to study other languages. Having grasped American English for launch, and British English before the end of 2016, Alexa has only added German since the start of 2018. Instead, Amazon have been broadening the range of devices in which Alexa might live.

ECHO SETUP

If you're thinking of joining the tens of millions who have purchased the iconic Echo speaker, here's the setup process:

Check your Amazon account details (typically an email and a password). You will also need to know the Wi-Fi password for your home.

Install the "Amazon Alexa" app on your phone. (Oddly, this is categorized as a Music app in the Apple App Store!) You need to go to the app's Settings page via the menu to add a new Echo.

Log in to the app, then plug in and turn on your Echo. It will generate a new Wi-Fi signal for you.

Now switch out of the Alexa app and use your phone's Settings page to connect to the Echo's Wi-Fi signal. The devices will now be able to exchange information. It will ask you which Wi-Fi network you want to use and what the password is. It is important you use the same network any other smart devices are connected to (probably your normal home network).

Now follow the on-screen instructions, and remember to reconnect your phone to your normal Wi-Fi network.

ASK ECHO SOMETHING, STARTING "ALEXA..."

THE APPROPRIATE RESPONSE IS SENT BACK

YOUR QUERY IS SENT VIA THE INTERNET TO AMAZON'S SYSTEM TO PROCESS

YOUR ECHO CARRIES OUT THE COMMAND

ECHO SHOW

The first but not the last of Amazon's range of Echo devices with screens (and cameras). With a display like this, the Show can even be placed in the kitchen to provide hands-free recipes to follow.

+

- **OUTSTANDING COMPATIBILITY WITH SMART HOME PRODUCTS**
- **GOOD UNDERSTANDING OF NATURAL LANGUAGE**
- **GOOD VARIETY OF INSTALLATION OPTIONS (AND PRICES)**

–

- **SETUP REQUIRES ANOTHER DEVICE**
- **PRIVACY CONCERNS MIGHT BE TROUBLING FOR SOME**
- **NEEDS A GOOD WI-FI SIGNAL AND INTERNET CONNECTION**
- **NOT A HUB (EXCEPT ECHO PLUS)**

SIRI & HOMEKIT

A leader in the pocket, but in the home, too?

The tech industry is fixated on Apple's every move, and understandably so. The smartphone revolution, iPods, and (for those with longer memories) even the idea of computer mice rather than typed commands were all brought to the world's attention — if not actually invented — by Apple. But what about the smart home? Is Siri a player in this new world?

This is a complicated question, because Apple have clearly chosen to play a different game to Amazon. While Amazon are happy making Alexa available to all and sundry, Apple are perhaps more interested in their own ecosystem. The iPhone creators have long offered Siri, the voice assistant, as a feature of the iPhone, however, going into 2018, they were already some way behind their rivals when the date for the release of their answer to Amazon's Echo, the HomePod, was moved again.

Apple, however, don't play in the lower end of the market price-wise, so it's not at all clear whether they viewed this as a problem. Their sophisticated HomePod puts audio quality first, and is considerably more expensive than the inexpensive entry offerings from Amazon and Google. Seemingly the plan is not to compete in this space, but to become a worry for high-end speaker manufacturers. Apple has always done well from music, so why not now?

If the HomePod is first of all a fancy speaker, then perhaps Apple's view (and it's far from an irrational one) is that we'll prefer to use the device in our pockets to command our smart home. Apple's HomeKit launched in 2014 as a way to allow the developers of smart home products to implement controls for iOS devices (iPhone and iPad). In turn, Apple have made these third-party devices visible to their products, so you don't always need to hunt for an app to — for example — switch on the lights; you can simply use Apple's "Control Center" page rather than locate the specific app.

Unfortunately, while this sounds great, Control Center is currently, at best, very wasteful of screen space; in iOS 11 you're only shown nine "favorite" devices at once and have to scroll inside a square window, despite the fact that acres of the screen are effectively blank.

Grumbles aside, Apple have built all the right tools for developers and, thanks to the iPhone's popularity (especially among those who define trends in tech), have a very widely supported system. It's virtually unthinkable to create a smart device without an iOS-compatible app to control it, which will make Apple important just as long as iPhones are.

IOS CONTROL

Apple's iOS detects HomeKit devices; your HomePod (above) will find these too.

+

- SOLID COMPATIBILITY
- GOOD UNDERSTANDING OF NATURAL LANGUAGE
- APPLE HAVE MADE IT EASIER FOR DEVELOPERS TO ADD IOS-FRIENDLY CONTROLS TO THEIR DEVICES

–

- NOT CHEAP
- APPLE'S INTEREST IN HOME AUTOMATION ISN'T CLEAR
- NOT A HUB
- NO REAL EQUIVALENT TO AMAZON'S SKILLS

GOOGLE HOME

You ask Google a lot of questions already, so why not a few more?

By the end of 2017, electronics retailers had just as many of Google's smart speakers as the Amazon ones. At this point you might have thought that, given its ubiquity, Google's success was guaranteed. It has to be said that, despite undoubtedly having the resources to dominate, their first moves in this area were mixed.

Google have now shrugged off what many see as a poor start in the field of digital assistants – the Google Glass. As well as being far from a commercial success, there were many concerns about the "creepyness" of the device, which incorporated a camera and a prism in headwear a bit like glasses, which relayed a picture to the user's eye.

Possibly, it was just a matter of timing; perhaps at the time of the Glass's launch (all the way back in 2012) the world simply wasn't ready. This, remember, was before Alexa had met the public, so talking to computers was pretty weird anyway. Inevitably, the Glass went away for a rethink, drifting from the public eye but still being applied in numerous ways. Lately, scientists at Stanford University have been conducting an experiment into how these glasses might help children with autism, for example.

The command "OK Google" has had a life beyond the Glass, though: the Google Assistant can be invoked on every Android phone (beyond version 4.4), like Siri on an Apple phone, and Google have put a good deal of energy into continuing to improve its functionality and intelligence. That is of real interest in the smart home world, since it is, of course, backed up with the ability for developers to create commands. Given the popularity of the Android phone platform, compatibility with Google technology is very desirable for smart-device manufacturers, too.

Already an experienced and powerful player in the software field, with the advantage of living inside a huge proportion of phones regardless of brand, Google decided to explore the hardware market. Since then, they've quickly become a significant player, with the Pixel-branded smartphone and their

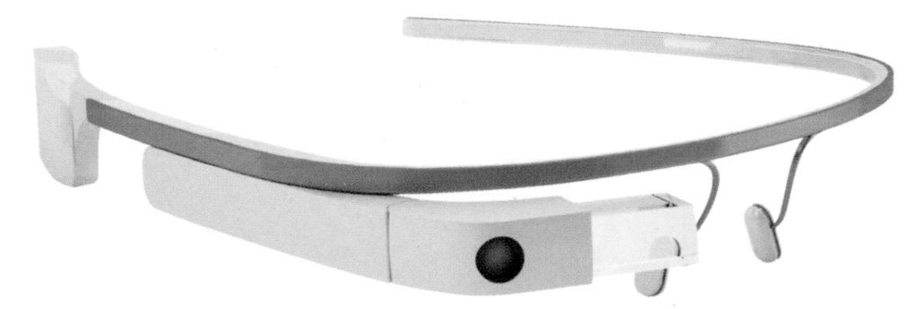

GOOGLE GLASS

It's widely accepted that Google went a little too far with its first foray into digital assistants.

GOOGLE HOME MINI

The equivalent of Amazon's Dot, this little device was selling for as little as $29 in the 2017 gifting season, and moving over six million units.

OK Google, make everything purple.

This command will work with smart lightbulbs.

GOOGLE PIXEL

With the release of the Pixel series, Google established themselves as a significant player in the hardware as well as software areas of the phone market. Just as Apple make their own operating system for their phone, so too do Google. The difference is that Google also license theirs to nearly every other phone manufacturer.

GOOGLE HOME

own smart speakers. The Pixel phone is important because the high-end phones are the ones that make the most money, and in the past, Google had left other companies, notably Samsung, to put a physical shine on their Android operating system.

The speakers (including the low-priced Mini) — branded Google Home — may be even more a signal of intent, since these are so clearly aimed at the same market as Amazon's Echo, with similar size, functionality, and pricing.

Alexa has Skills (more than 10,000 at the last count, though many of them are quite silly). Google offer the ability to extend their platform with "Actions." At the time of writing there are considerably fewer of these, and even if the number grows rapidly, there is a quality-versus-quantity argument that makes comparison tricky. Assuming Google behave as they do with the Android app store, one wouldn't expect them to be too picky, though.

Google, of course, do bring something else to the table: their extensive online services (Gmail, Calendar, Google Drive) which should, in theory, lead to more advanced integration.

Like Amazon, Google have taken the view that the Google Assistant technology, which processes the voice commands, can be licensed to anyone who wants to add it to their own devices, making the prospect of repeating the success of Android all the higher.

GOOGLE HOME

First released in the US in 2016, two years after Amazon's first Echo, Google's smart speaker aims squarely at its rival's market.

One note, of course, is that, as with all such technologies, once the key phrase has been detected by your device (phone, smart speaker etc.) the processing happens on powerful remote computers. That means Google are potentially storing your data and using it both for good (improving their software) and for other purposes that you might not feel so comfortable about (like identifying your behavior patterns to predict your needs).

CES 2018, LAS VEGAS

At the world's largest electronics trade show – where Google had never been a major player before – the company really made an entrance with their digital assistant brand in 2018, plastering it over the host city's advertising locations.

+

- **GREAT UNDERSTANDING OF NATURAL LANGUAGE**
- **CHROMECAST MULTI-ROOM AUDIO SYSTEM**

–

- **PRIVACY CONCERNS MIGHT BE TROUBLING FOR SOME**
- **NEEDS A GOOD WI-FI SIGNAL AND INTERNET CONNECTION**
- **NOT A HUB**

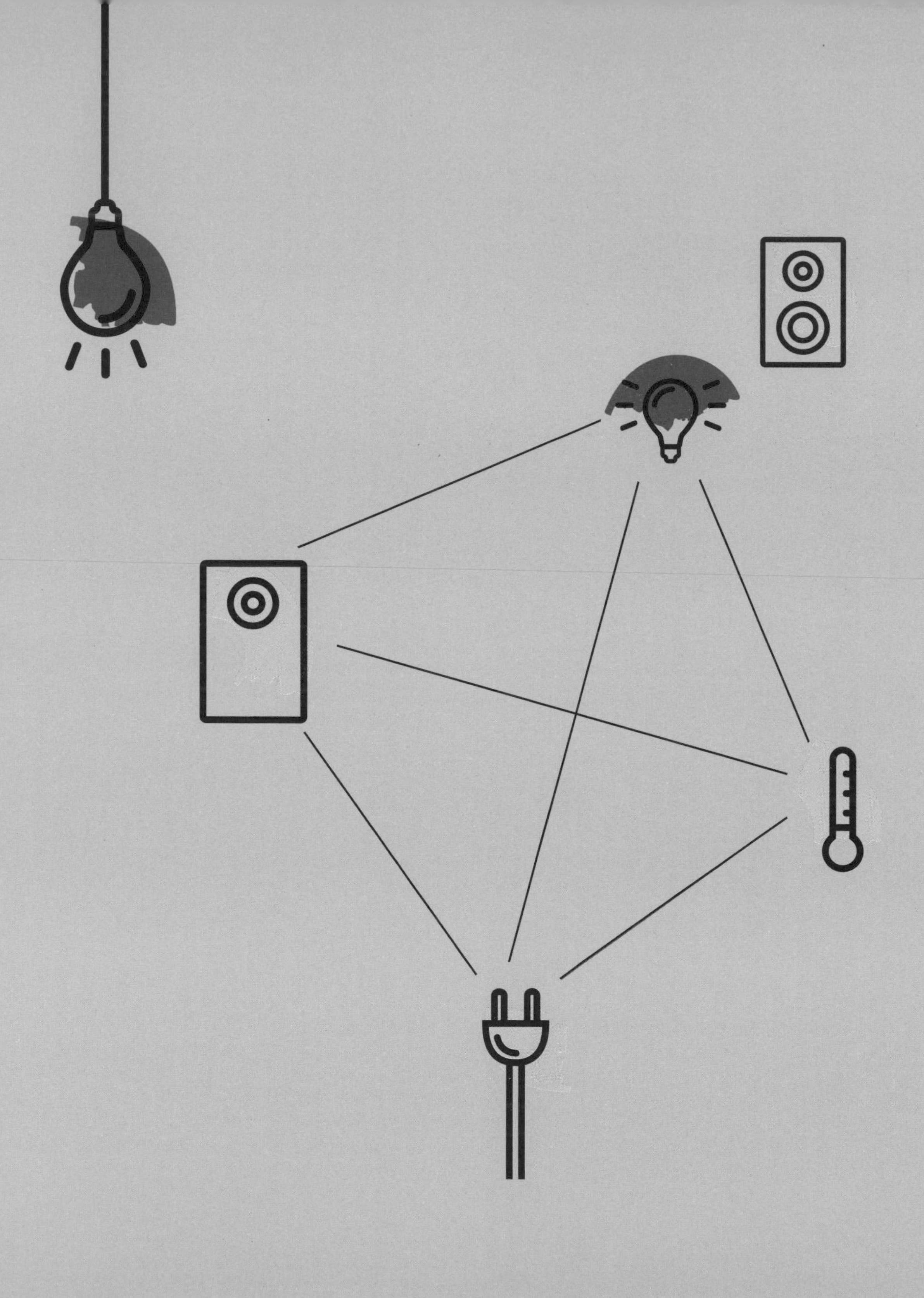

SMART HOME SYSTEMS

Outside the home, you reach smart tech via the internet, apps, and technology we're well used to. This chapter is about how the new Internet of Things joins and adds to that world.

HOW YOUR INTERNET WORKS

Understanding where the internet and your network meet

In its simplest terms, the internet works by directing small bits of information to their destination. Even though you don't usually know, or need to know, what it is, every connected computer has its own address. This is called the IP (Internet Protocol) address, traditionally written in the form "000.000.000.000."

When you're using the web, a service called the Domain Name Service (DNS) acts like a phone book for your computer, translating a (memorable) web address, such as www.tamesky.com, into the (all but impossible to remember) numerical IP address that computers understand.

Your internet connection is managed by your Internet Service Provider (ISP), which sends your request to a hierarchy of router servers above them until it achieves its destination. The data that makes up the page will then be sent back over the internet, each packet traveling along a route (and redirected by a number of servers on the way), to your computer. These servers are able to reroute the data should one fail.

Internet Protocol is how "packets" of data are directed over the internet between the computer making the request and the location of the requested information. It works with something called Transmission Control Protocol (TCP), which, in its simplest terms, takes the information from the internet and sends it to the correct application on your computer, checking that it's OK and asking for replacement data packets if there are issues. Together this makes up the TCP/IP.

This system of moving data has changed the world. Significantly, the routers further up the chain simply move the data they're given regardless of its source, which has meant that a number of new technologies have been able to establish themselves without having to build their own network first—YouTube and Netflix being notable examples.

It is technically possible for your Internet Service Provider to monitor the source of your incoming data and make it harder for some information to get through. This is particularly significant if, for example, your service provider is also your cable TV company and they decide they'd rather you used their TV service than Netflix. It's definitely worth checking the policy before you commit.

Inside the home, your networks will also use TCP/IP naming conventions, but these will likely be closed networks, so the local addresses are not necessarily shared with the world.

Your modem routes data to (and from) all the devices connected to your home networks, including those that use Wi-Fi. Other networks your smart devices might use (such as Zigbee Z-Wave, and Thread; see pages 34–39) will be linked to your modem via a hub, which essentially acts as a translator between the languages they speak and the protocols the internet works on.

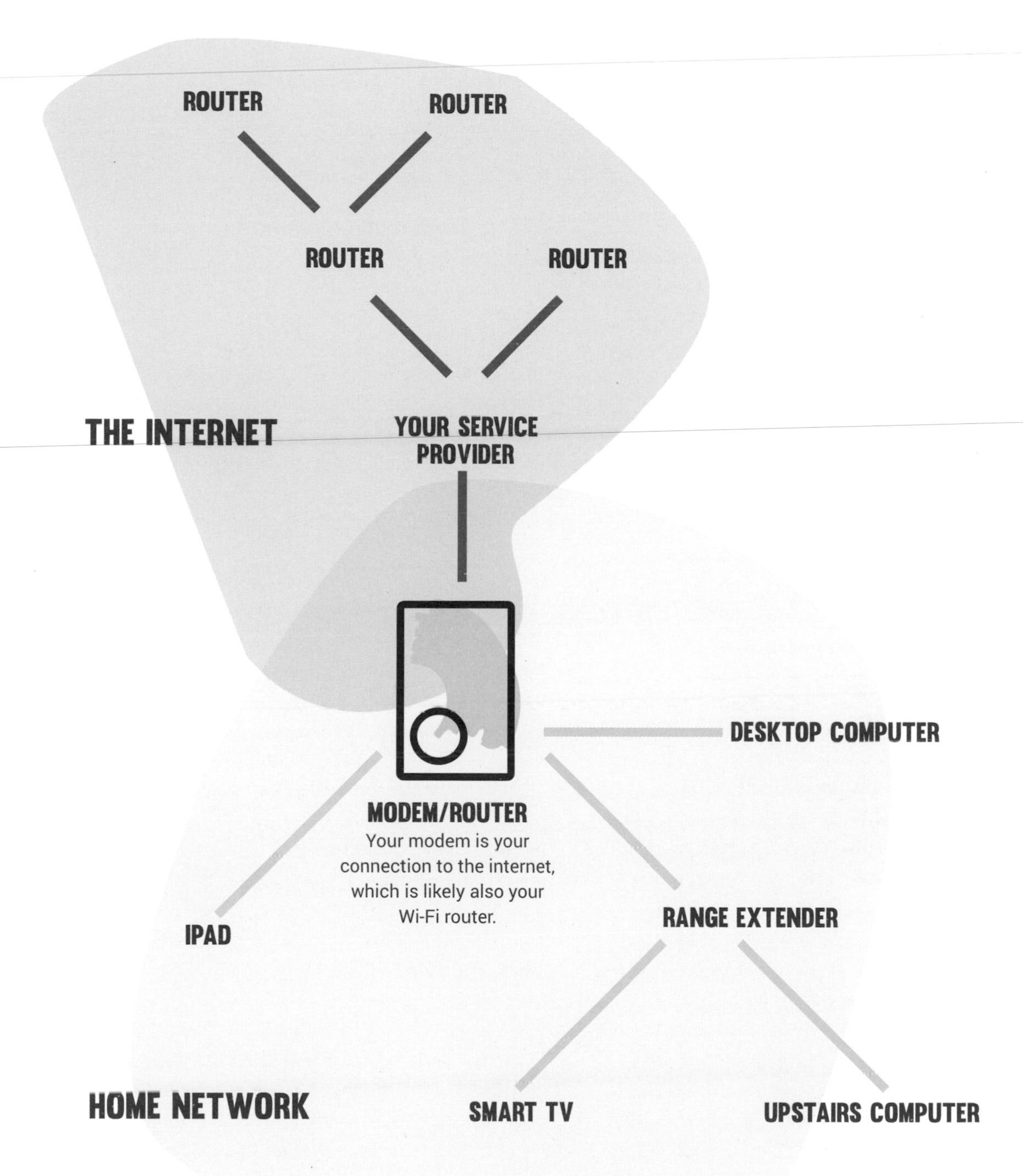
ROUTER
ROUTER
ROUTER
ROUTER
THE INTERNET
YOUR SERVICE PROVIDER
DESKTOP COMPUTER
MODEM/ROUTER
Your modem is your connection to the internet, which is likely also your Wi-Fi router.
IPAD
RANGE EXTENDER
HOME NETWORK
SMART TV
UPSTAIRS COMPUTER

MESH NETWORKS

A wireless technology designed for the smart home

The reason it's worth knowing a bit about how the internet works is because it's not the only network that will be involved in your smart home. Depending on what you're setting up, you'll likely hear about a number of other networking technologies.

The best known network is, of course, Wi-Fi. I say "of course," but most people don't know that Wi-Fi and the internet are two different things. Wi-Fi is actually a separate network that allows devices to communicate with each other and/or with the internet – so your computer could communicate with your printer over Wi-Fi, without being connected to the internet. The reason for the confusion is probably that the capacity for Wi-Fi is often housed in your normal router, so you're used to seeing Wi-Fi as an extension of the internet.

Well-known examples of other networking technology within the smart world include Zigbee and Z-Wave. These are both "mesh network" systems, meaning they essentially use all the compatible devices in your home to build their own mini-internet, each acting as a relay station to pass the message or command on to the relevant device. This is useful because it resolves a common problem that Wi-Fi users experience: Wi-Fi tends to have issues with range and transmission in even modest homes. You'll often find that as you get physically closer to the router, the quality of your connection improves, enabling faster and more reliable data transfer.

The other issue with Wi-Fi is that it has relatively high power-consumption needs, which means that it works best with devices connected to the mains supply. The trend is to move away from that, though, and many devices created for mesh networks use far less energy, requiring only a small battery for years of use. Smart home devices compete to be easier to install and more convenient in operation – the difference between a switch that you can stick on the wall and one you need to wire in is obvious.

Acting as a point on a mesh network relaying data does mean a bit of extra work and a bit of extra power usage, so in some cases mesh networks deliberately stick to using those devices with outlet/mains power connections to carry that load, and leave the battery-powered devices to operate only when they need to. All of this happens automatically.

COIN-CELL POWER

Many Zigbee/Z-Wave devices can be powered by a single coin-cell battery for around two years; Wi-Fi would manage days only.

WI-FI — A STAR NETWORK

On a star network — like Wi-Fi — all connected points communicate via their hub. That means that even if you're in the same room as the device you want to control, your command first has to go from, say, your tablet, to whatever room the router is in, then back to the device in question.

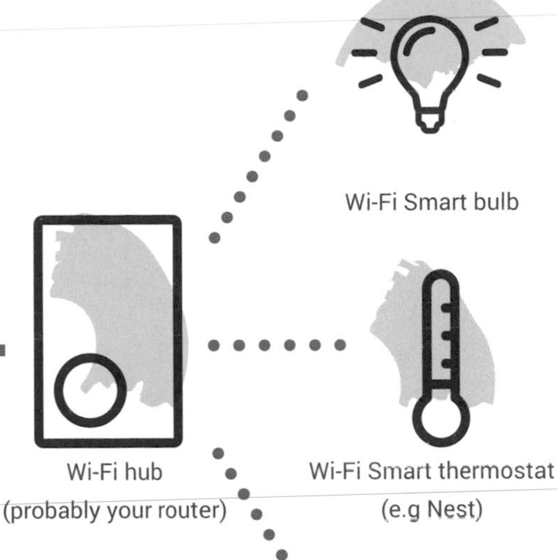

A MESH NETWORK

A mesh network operates as an interconnected web, each point of which generates a signal, so that even when devices are far from the hub, if they're close to another mesh device, the command gets through.

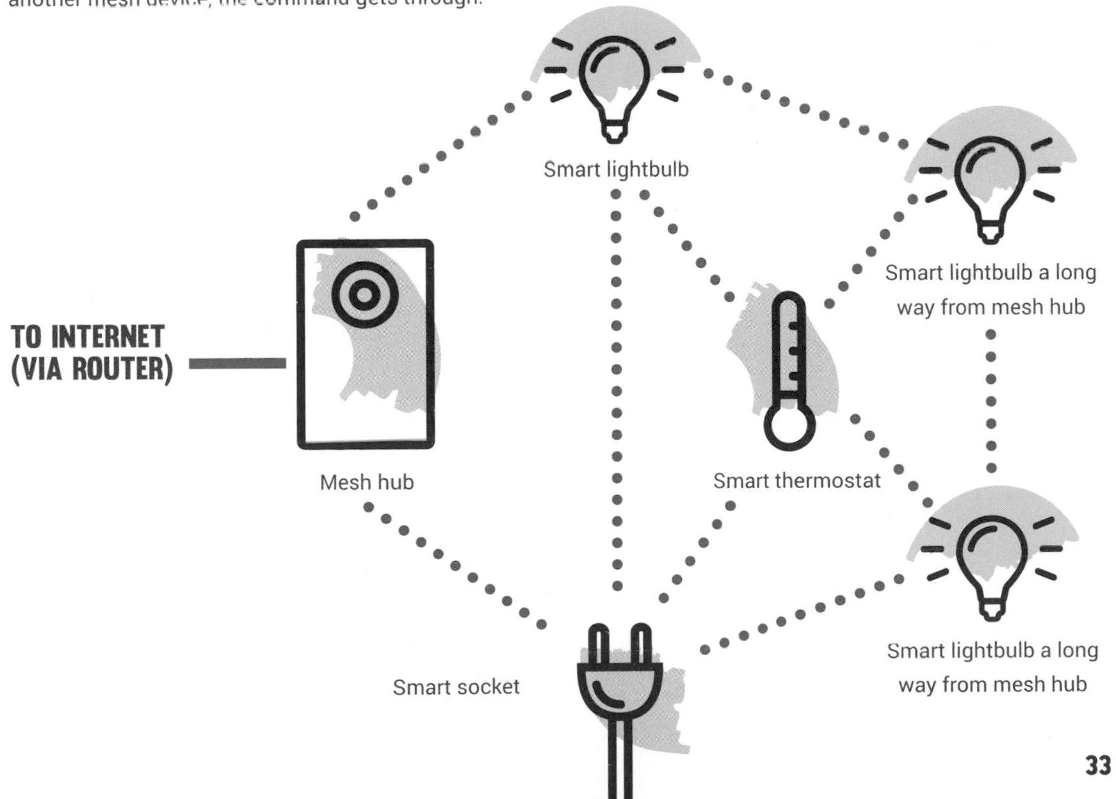

ZIGBEE

The open smart home alliance

Zigbee is a type of wireless network (or "standard") found in a broad range of smart home products. It has built-in encryption for security and is used in a lot of different products (LG, GE, Logitech, and plenty of other brands use Zigbee). It doesn't require much by way of setup by the user.

Despite finding themselves at the same point in the alphabet, and sharing many technical aspects, Zigbee is not the same as, or compatible with, Z-Wave (which will be introduced overleaf).

Zigbee has been around for a while now; it was first set as a standard in 2003, and since then there have been separate generations, known as Zigbee Light Link (ZLL) and Zigbee Home Automation (ZHA). These have been brought together as Zigbee 3.0.

It is an open standard, based on (but not the same as) the snappily named IEEE 802.15.4. IEEE stands for Institute of Electrical and Electronics Engineers; believe it or not, you'll be familiar with some of their other output: 802.11, for instance, is better known as Wi-Fi, and 802.15.1 is Bluetooth. Zigbee, however, is a different standard and as such you'll need a hub for them to communicate. Nonetheless, Zigbee operates on a radio standard, which means that it's easy to add the protocol to existing components.

Zigbee creates a mesh network that's ideal for use in smart home devices. As an open standard it's also easily accessible for a wide range of manufacturers (and entirely free for academics and enthusiast developers). All that is definitely encouraging from the perspective of price-reducing competition, although in the past it was known to suffer from interoperability issues. Zigbee can, however, also be used as the basis of a closed system, such as the Philips Hue lighting system.

Finally, Zigbee is capable of working on both the 2.4 GHz radio standard (also used by Wi-Fi) and at the slower 915 MHz (US) or 868 MHz (EU); having two options means that the system need not be affected by (or affect) domestic Wi-Fi, and gives it an alternative when it comes to traveling through walls.

- **MESH NETWORK**
- **UP TO 65,000 DEVICES ON A NETWORK**
- **40FT/10M RANGE (UNOBSTRUCTED, BETWEEN DEVICES)**
- **40–250 KBPS**
- **BUILT-IN 128-BIT SECURITY**
- **TWO-SPEED RADIO**

ZIGBEE HUB

There are numerous hubs available incorporating Zigbee, including this one from Orvibo.

AMAZON ECHO PLUS

It is a mark of the success of the Zigbee standard that Amazon have created a version of the Echo that includes Zigbee compatibility.

IHOMEWARE SMOKE DETECTOR

This security device is Zigbee-based, and can be linked to other smart devices so it will go off when any of its connected sensors are tripped.

Z-WAVE

One of the tech world's leading smart home alliances

Optimized for battery-powered devices, the system defaults to a power-saving mode and has been widely adopted by product creators — including well over 600 certified manufacturers — and used in over 2,000 devices. It is simple to set up, and all devices displaying the Z-Wave logo should be compatible with each other, even when there's a big age gap between them.

Like Zigbee, Z-Wave is a mesh network capable of linking devices in a chain that can automatically send signals via each other (this is known as "hopping") in order to ensure a better chance of a command getting through to a device. Like Zigbee, it uses 128-bit security — a very good level of encryption. Unlike Zigbee, Z-Wave is not an open standard, meaning that to sell Z-Wave products, manufacturers need to pay for a license and follow the standard. The result is that interoperability is guaranteed.

Z-Wave devices also have home regions — rather like DVDs — and the reason behind this is that Z-Wave devices only have one radio frequency (in the 800-900MHz band) which varies between different countries, while those with Zigbee can also use the universal 2.4 GHz one.

That also means that, again like Zigbee, your home Wi-Fi network isn't automatically compatible with these devices without some kind of hub, although as that generally is the case with most smart home devices, that isn't a big issue.

MYHOMEBOX

A Z-Wave-compatible device that can be set up to control one or a group of devices. What could be easier to use than a simple push button?

- **MESH NETWORK**
- **UP TO 232 DEVICES ON A NETWORK**
- **100FT/30M RANGE (UNOBSTRUCTED, BETWEEN DEVICES)**
- **40-100 KBPS**
- **BUILT-IN 128-BIT SECURITY**

Z-WAVE MEMBERS

The Z-Wave Alliance at CES, the world's largest electronics trade show. As you can see, numerous other brands form part of the booth.

Z-WAVE BULB

This generic smart bulb is Z-Wave compatible.

MYHOMEBOX

The MYHOMEBOX is a Z-Wave hub.

THREAD

A future-proof system for smart home technology?

A relative youth, Thread was announced in 2014 and it uses the same basic hardware as Zigbee. It was originally introduced as a cooperation between Samsung, Nest Labs, and ARM, with an eye toward the future of the Internet of Things.

While, writ large, it runs on the same standard as Zigbee (that catchy IEEE 802.15.4), Thread's network specification is based on IPv6 – a new-generation version of Internet Protocol – making it easy for companies to develop new products, because Internet Protocol is, of course, well established and well supported.

Since one of the partners behind Thread is Nest Labs, now owned by Alphabet (who are, significantly, also Google's parent company), there might be a little bit of trouble persuading other companies to adopt the standard, depending on their feelings about Google or the other partners.

On the plus side, Thread can be adopted as a software update in many existing products, thanks to the IEEE 802.15.4 underpinnings it shares with Zigbee. In fact, Zigbee and Thread seem to have become very close; a recent collaboration has seen Zigbee sharing its application layer with Thread, meaning that Zigbee devices can operate over Thread's network. For technical reasons, Thread can't work on Zigbee, so it would have been understandable if Zigbee wasn't too keen on this partnership, yet Thread has gained a credibility among device-makers and Zigbee do seem to be enthusiastic partners.

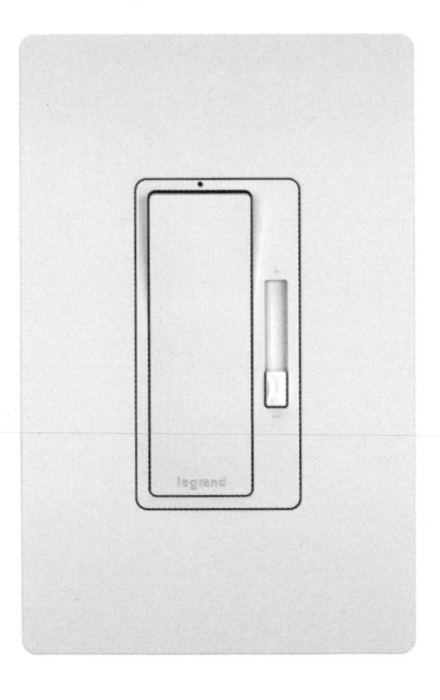

LEGRAND RADIANT

Legrand's range of light switches are based on the Thread system.

- **MESH NETWORK**
- **UP TO 250 DEVICES ON A NETWORK**
- **100FT/30M RANGE (BETWEEN DEVICES)**
- **40-250 KBPS**
- **BUILT-IN TWO-LAYER SECURITY**

NEST

Nest's home security system is compatible with Thread, but some of their older devices are not.

BLUETOOTH MESH

Announced in 2017, Bluetooth Mesh is an extension of Bluetooth Low Energy (the standard used by many wireless headphones). It adds the ability to use a mesh network with around 330ft (100m) of range, 32,000 devices, and a speedy 1 Mbps of data transfer (depending on the device).

Given the history of the Bluetooth brand, it's easy to imagine their mesh system doing well; for now there is little I can do except tell you to keep your eyes peeled for it.

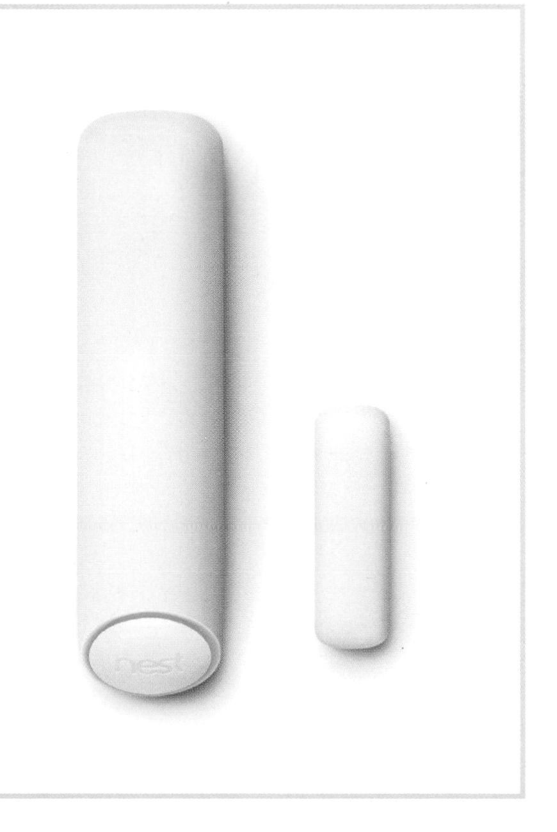

SMARTTHINGS

Samsung do smart homes, hub style

So far, we've mostly dealt with the best-known aspect of the smart home: the speakers and digital assistants that listen for commands. Most of these are not hubs, though, and since many of the smart devices run on standards like Zigbee, Z-Wave, and Thread, they may not be compatible with your existing technology without another device in between.

Hubs are the devices that connect to your existing home network (or Wi-Fi) and translate your commands into the signal that the smart device is looking for. So if one device (or set of devices) uses one signal — say you've got some lights that operate over Zigbee — and your "traditional" home network another, you need a hub to make them speak to each other. Hubs are the smart home's dirty little secret, never mentioned in the glossy advertising because an anonymous box that'll cost you extra money doesn't really fit with the slick "Computer, turn my room purple" commands you might see on TV.

Samsung, however, didn't make their voice assistant, Bixby, the centerpiece of their initial smart home offering; they opened with a simple hub, a product they bought from a Kickstarter startup and made their own.

The result is a home automation system that prioritizes practicality and interoperability between the devices under its SmartThings brand. By putting a cohesive system of products first, before engaging in the popularity contest of the charismatic assistants, SmartThings benefits from increased functionality. These devices can be programmed to perform clever responses to events; such as the lights coming on as you walk up your drive to your security system activating and your thermostat turning the heating off as you leave your house.

This, at least, is its heritage, at least. IoT (and the humble Bixby) is now a big part of Samsung's plans, which, in theory, should give you a practical underpinning and all the sci-fi fun available, if you want it.

If you're a Samsung fan, you might have spotted that other brands, like Samsung Connect and Samsung Smart Home, have been used for Samsung's smart home tech; they even had one kind of voice recognition on their watches and a different version on their phones. As of the beginning of 2018, the company announced that its 40 different apps will be consolidated into SmartThings.

The current SmartThings Hub supports Zigbee and Z-Wave, allowing Samsung to take advantage of the benefits of these standards, including mesh networking and lower-power function for appropriate devices. Quite smartly, Samsung are also offering the hub functionality into a number of other devices, including a SmartThings Link USB stick for the NVIDIA Shield TV (see page 49) and some of their home products, such as refrigerators.

SMARTHINGS HUB

Samsung's hub is compatible with Amazon Alexa and Google Home, so you can operate your devices via your preferred platform (unless it's Apple).

SMART FRIDGE

Building a SmartThings Hub into the fridge makes it instantly able to take over the "hub of the home" role.

SMART LINK

Samsung's SmartThings Link USB stick connects with NVIDIA's Shield, allowing you to add Zigbee and Z-Wave compatibility to your favorite smart TV system.

ENTERTAINMENT

For many people, the whole point of adopting new smart technology and revolutionizing their home has been for the sake of entertainment.

THE REAL HOME HUB

One piece of tech has long served as the home's centerpiece

Back in the 1950s, the television stopped being a niche technical showpiece and took its place at the heart of family life. By 1955, more than 50 percent of American households had a television set, and since then they've never gone away (less than 2 percent had held out at the turn of the century).

Television has been through several distinct stages: black-and-white, color, flat-panels, high definition (HD), and now smart TV. HD brought with it digital transmission; at that point television and computers looked at video the same way—as a series of ones and zeros encoded for transmission, and decoded at the other end into the moving images we expect to see.

From the point they started sharing a language, it was inevitable that the world of television and the internet would collide. And now it absolutely has done, with many new televisions featuring some kind of smart technology built in. The most popular system is Android TV, a variant of Google's phone-operating system that the tech giant has licensed to Sony, Sharp, Philips, and other manufacturers to build in to their sets.

Just as in the phone-operating system, apps can be added to these televisions, and manufacturers are sure to include popular streaming services like Netflix by default. If you've used a set like this, you'll know that it's great to have the ability to stream high-resolution video to your TV.

As the heart of your smart home, the TV does, however, have some restrictions. There are limits to the use of a TV remote control when it comes to other tech around the home, and the operation of a TV is quite linear; you wouldn't want to have to use the TV to dim the lights because you'd have to switch from what you were viewing (say, in the Netflix app) back to the menu, and then go to the light-dimming app (before making your way back to Netflix).

Where smartness in televisions is most successful is in how it simplifies the TV experience (and access to content). There are no more VHS cassettes or DVDs to burn. Even if you are watching a traditional TV channel, with a smart TV it's entirely likely that you can still press pause and use the power of digital to store the stream of data until you've finished making that coffee. For many, TiVo first introduced this concept.

Ironically, with all that power, we actually lose a certain amount of control over "our" content. For the most part, digital TV is stored on remote servers. You either buy access through a streaming service like Netflix or Amazon Prime (where shows or movies are available for a certain time, then may vanish), or one like Apple's iTunes Store, which allows you to buy the movie, so it'll be part of your collection for all time, but can only be accessed via Apple and through the specific Apple ID that purchased it.

In both cases you are dependent on an internet connection and the company you've signed up with remaining in business and able to reach you.

There are some solutions to this. You can copy the content of your DVDs and other media onto a physical drive and link it to the television — many televisions have USB ports that could connect to a hard drive (or a USB stick) for exactly this purpose. A more sophisticated solution is to download an app that connects to a drive elsewhere on your home network.

BRAVIA TV

Modern televisions are now so thin that manufacturers need to look for other ways to impress consumers. This BRAVIA, for example, uses the flat panel itself as a speaker by vibrating the screen — as well, of course, as including Android internet TV services (there's a Netflix button on the remote).

SMARTENING YOUR TV

You don't need to buy a new television set for the smart tech experience

If you've (fairly) recently invested in a lovely HD TV, you may look enviously at the updated smart features on newer models, but there's no need to discard the whole set in order to add Android, Amazon, or even Apple functionality to the one you have.

There was a time when being able to view anything other than standard channels on your TV required adding a big box. The term for this was "set-top box," since your cable or satellite decoder could comfortably sit atop a television set. These days, though, televisions are thinner, so the term has fallen into disuse and the devices themselves have had to find a new position.

As TVs have gotten thinner and bigger, the boxes have shrunk and now tend to sit under or around the back of your TV. Unlike in a VCR, there are no moving parts, and the electronics required to display video are not dissimilar to those in current phones; the processing power to handle all the pixels (dots) that make up a 4K TV picture is broadly the same as that found in a higher-end mobile phone.

Apple, for example, produce a version of their mobile operating system, iOS, for televisions, which currently fits into a square black box about the size of a hockey puck. The tvOS (Apple TV's operating system) runs on very much the same electronics as an iPhone, and the main reason the box needs to be bigger is to incorporate the necessary electronics to use a outlet power.

The Apple TV connects to one of your television's HDMI sockets (like a Blu-ray player, satellite box, or similar) and comes with a small remote that lets you invoke Siri with commands like "Search YouTube for cat videos," or to search other apps for content. It can also receive commands for Apple's HomeKit; the remote control needs to be plugged into the same "Lightning" connector that the iPhone uses to charge. Apple are sometimes frustratingly keen on a unified ecosystem, but Apple TV's third generation and beyond have been excellent machines.

There is a lot of competition to control the TV (from a corporate perspective), and Amazon and Google have taken a slightly different approach, offering something even more discreet should you wish it: the Amazon Fire TV Stick and Google's Chromecast. Amazon's first effort was exactly what the name suggests: a small stick that plugs directly into your HDMI port and (hopefully) won't block any of the neighboring ports on your TV.

Google had the same idea with their first hardware effort, but their design consisted of a tiny device dangling on a normal-sized cable and plug, because that caused less interference with TV designs. Amazon Fire TV eventually followed Google's example for their Ultra model, with the dangling pendant design favored over the stick. If you prefer to use the stick design, however, you can get hold of a short HDMI extension lead with a socket at one end and a plug at the other.

FIRE TV STICK

Amazon Fire TV Stick and remote control; the remote can be used to navigate the menus manually, or spoken into.

HDMI EXTENSION

If your stick doesn't fit, an extension cable is a cheap solution.

Alexa, watch *Star Trek: Discovery*.

Depending where you live, it's possible to control most of the major apps on a Fire TV Stick using Alexa; Americans will find Bravo, CBS All Access, Hulu, NBC, Showtime, PlayStation Vue, Netflix, Prime Video, and HBO Now.

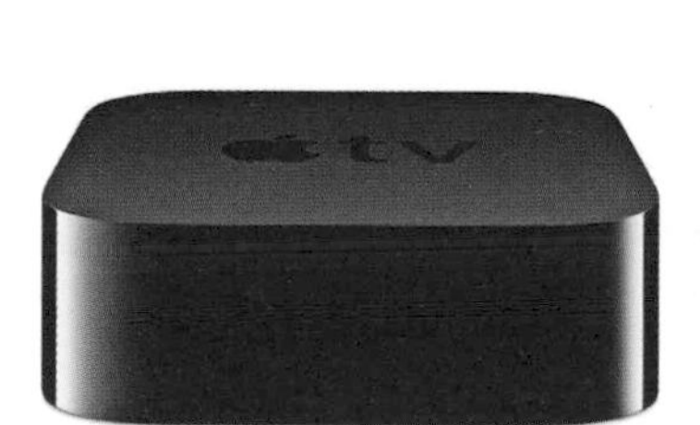

MENU

ROKU STICK

Amazon and Google aren't the only ones in the game. If anything, services like Netflix (note the button on the remote) have been the key drivers.

APPLE TV

Apple TV offers a discreet box that plugs into the wall, plus a sleek and typically pared-back remote control.

SMARTENING YOUR TV

Google's Chromecast isn't quite the same thing as Android TV; the former treats your phone as the principle mode of interaction, while Android TV (like Apple TV) uses a small remote. The Chromecast is essentially a tool for streaming content from your phone or tablet to your television screen via an app. It has very little onboard storage for additional apps like Netflix, just what is required to keep its own features working, so it's essentially relying on your phone or device to do all the hard work. The benefit of having no remote and a pared-down device is that it's cheap and easy to hide under your TV. The downside is that your continued viewing pleasure relies on your device not running out of batteries.

Screencasting is a feature that exists in Android TV and Apple TV (the latter, naturally, insisting on you using it only via other Apple products, while Google are rather more open) that lets you use your phone to control what's shown on the TV. It's known as screencasting, but the buttons that let you pause or choose the next thing to watch stay on your phone screen.

Most of these devices have been updated with more powerful processors since their first iterations, making for a smoother experience with menus and so on. In fact, this gives you something of an advantage over those who go for the TV with built-in streaming functionality; it's a lot more expensive to replace the whole set.

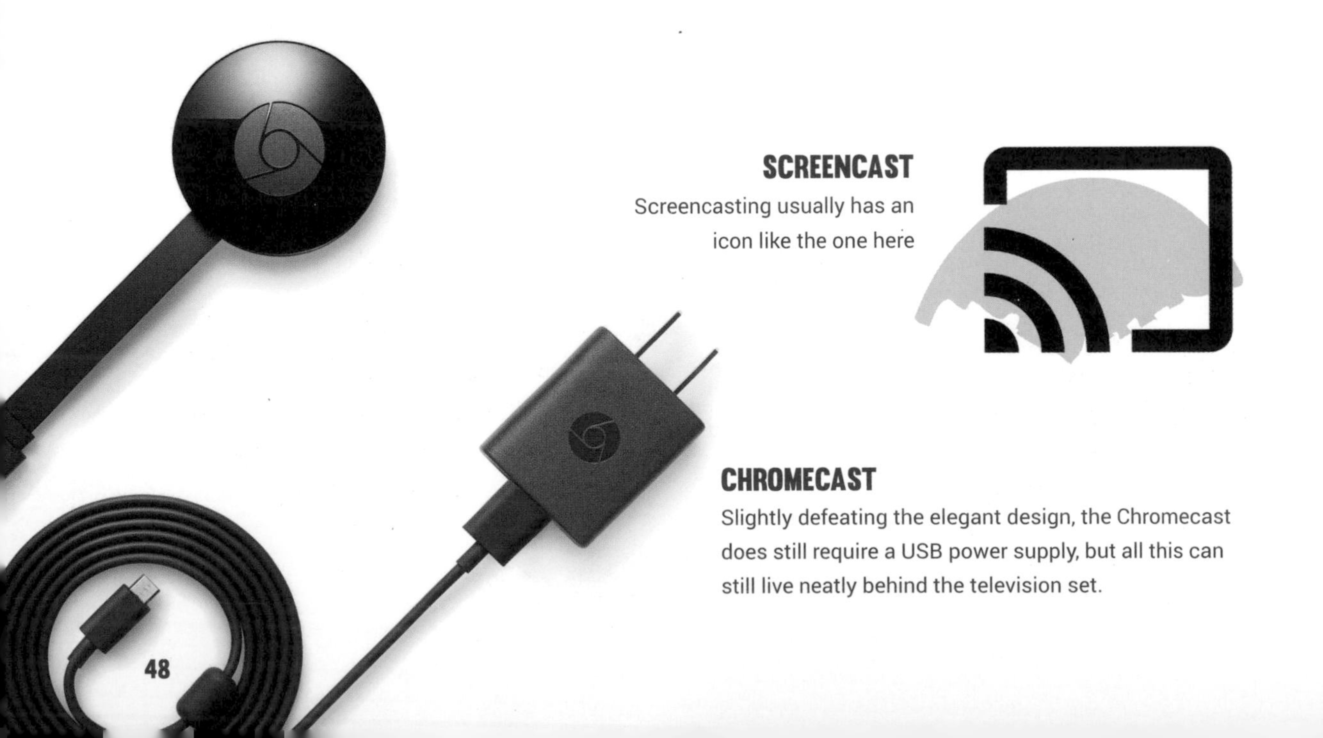

SCREENCAST
Screencasting usually has an icon like the one here

CHROMECAST
Slightly defeating the elegant design, the Chromecast does still require a USB power supply, but all this can still live neatly behind the television set.

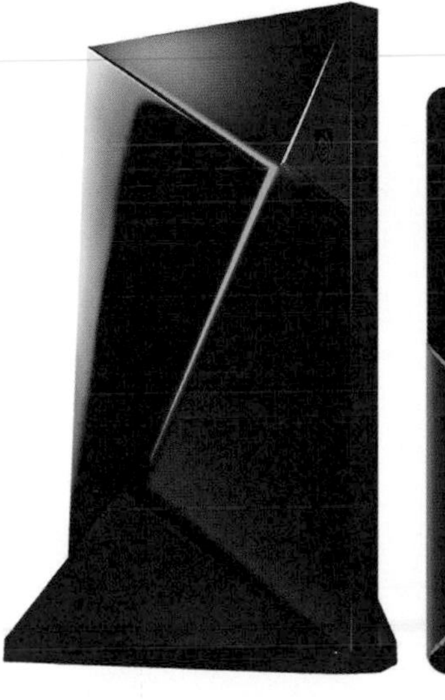

NVIDIA SHIELD

With built-in Android, and Google Assistant accessed via the mic in the remote, the Shield seems like many of the other devices here, but NVIDIA produce powerful graphics systems for PC gamers, and that heritage is built in. This device syncs games straight from a gaming PC to your TV screen, too. Samsung have teamed up with NVIDIA with the SmartThings Link USB, essentially allowing you to turn your TV into a hub (see page 40)

My partner has connected Apple TV to an oldish TV in the room where I do my sewing and it's great. It's easy to use and works on the same Netflix account. I don't understand why he didn't use the cheaper Amazon Fire TV Stick though.

+

- **MULTIPLE CONTENT PROVIDERS SUPPORTED**
- **SOME PROVIDERS' APPS ALLOW ACCESS TO LIVE TV (INCLUDING SPORT) AT A LOWER PRICE THAN VIA SATELLITE OR CABLE**
- **A DIGITAL ASSISTANT CAN FIND CONTENT FOR YOU**

−

- **STICKS AND DONGLES FEEL INELEGANT**
- **4K UHD HDR OFTEN REQUIRES A SPECIAL HIGHER-COST SUBSCRIPTION**
- **IF YOU BUY CONTENT, YOU ARE LOCKED INTO AN ECOSYSTEM**

SMART TV CONTENT

So where does all this content come from?

It can be a little confusing to navigate the world of digital TV offerings for so-called "cord-cutters" (those people breaking the connection to cable, satellite, or antenna), with an array of devices, streaming providers, live streaming services, and more.

This page is designed to give you an introduction to the possibilities and to broadly categorize the providers (not all are available in every country).

NETFLIX

A content provider with an app available on most platforms. Netflix made 6 billion dollars worth of video in 2017, with a projected $7 billion for 2018. It was spun-off from Roku in 2008.

Monthly subscription.

AMAZON VIDEO

Both hardware (Fire TV) and content creator (Amazon Studios), the latter investing $4.5 billion in new video in 2017. Amazon Prime also provides access to non-Amazon-originated content.

Monthly subscription (Amazon Prime) and one-off purchases.

VUE

Under Sony's PlayStation brand (though not requiring a PlayStation) this is an over-the-top streaming service (see Sling TV, right) with on-demand options.

Monthly subscription with different plans.

SLING TV

Known as an "over-the-top service"—meaning it allows access to certain TV channels via the internet for a subscription fee that you pay in addition to your internet costs.

Monthly subscription with different plans.

APPLE

Maker of hardware that can run Amazon, Netflix, and other apps, but rumored to have invested $1 billion in original content.

Device and one-off purchases of content (monthly subscription for music only).

ROKU

Hardware creator that makes video-streaming devices for televisions, allowing access to a variety of TV channels, including Netflix and Amazon Video.

Fees according to content providers.

TV NOW

Today's connected TVs can give you access to a huge range of TV channels, far beyond the selection featured here. You can also enter apps more usually associated with dedicated computers or smartphones, like Facebook and YouTube.

NOW TV

Essentially the digital streaming version of Sky TV (the UK satellite broadcaster). Movies, entertainment, and sports packages are available, and UK customers will find HBO originals here.

Monthly subscription.

HBO

Part of Time Warner Inc., this is a premium channel in the US, spending a bit over $2 billion annually on content. HBO GO and HBO NOW are its streaming brands.

Monthly subscription.

HULU PLUS

A Fox/Disney/NBC enterprise originally created as an over-the-top service, but now investing $2.5 billion a year in original programming.

Monthly subscription.

MULTI-ROOM SOUND

Get ready to dance through your home

The idea of multi-room sound once meant having a system built in, not unlike the school announcer. More recently, it's been possible to use networking to make the idea of drifting from room to room listening to consistent sounds a reality.

Now there are a number of relatively easy-to-install systems available that can piggyback onto your home system. That's great for a number of reasons, not least because, for most people, music streamed from an online service has replaced physical sources such as CDs and vinyl.

Listening to music is most practical if you use one of the streaming services — Spotify, Apple Music, Tidal, Google Play Music, Amazon Music Unlimited, or Napster. Of these options, Spotify is the most established provider and it is supported by every speaker system on the market (except Apple's).

In terms of systems that can link speakers, many that are available are closed systems. Sonos, for example, is a big brand in the field, which creates a proprietary mesh network between its own devices that's incompatible with others (although it uses your Wi-Fi to stream its audio).

Predictably, the alternatives tend to come from certain brands that will be becoming rather familiar. Apple offer AirPlay 2, which is compatible with a number of brands (including Sonos), while Google's Chromecast is built into many devices or available with Chromecast Audio. Both systems have now been made compatible with a greater range of brands and, using their respective apps, can direct music to a particular room and "group" of speakers, so that when you play music from the app, it is sent to all the speakers at once.

CHROMECAST AUDIO

Similar to the Chromecast for TV, this inexpensive gadget can stream audio to a wide range of devices.

ON THE 45

The idea of a physical music collection is becoming a thing of the past.

SONOS ONE

One of Sonos's range of speakers has added digital assistant features.

SONOS CONNECT

Link Hi-Fi or home cinema to your Sonos system.

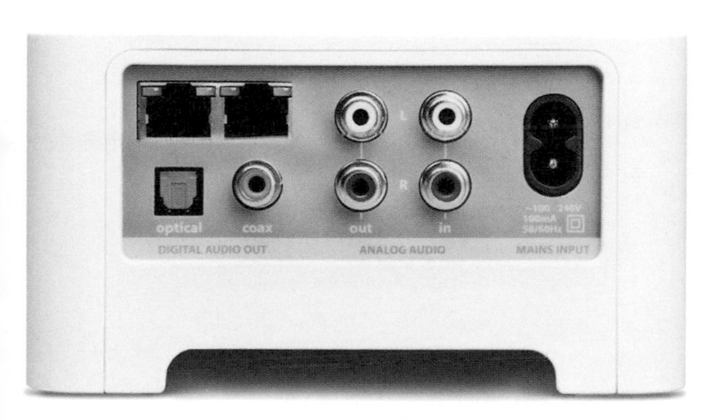

INDIVIDUALS & FAMILIES

Shared lists and privacy

One thing that you tend not to appreciate until it's gone is the privacy surrounding your personal media. There are forms of content that people want to keep private for obvious reasons, but it's likely you own a few CDs or records you both love and feel slightly ashamed of. What happens to them?

The last thing any of us want is for our guilty pleasures to be exposed to our family, friends, or housemates, but how exactly is this kind of information protected when one can't simply hide the shameful "Titanic Soundtrack" CD behind the wardrobe, and leave all the "cool" stuff face out in the living room?

At first digital music seemed to make it all the easier to build up a collection of exactly what you wanted, no matter society's public views on Coldplay. It seems likely that one of the contributing factors to the success of the iPod that reassuring "i" in the name, which, along with the presence of headphones, reinforced the idea of a personal collection. Other music or media platforms are equally or even more privacy-conscious; you tend to need a password to access a laptop if that's where your iTunes or Spotify account resides. And all of these handle our privacy using the login (password and email, for example), an approach we've become used to.

You will likely retain some privacy around your media choices by using your individual mobile phones as portable music players, but what about your smart speakers and Smart TV devices? They're available to everyone who walks through the house, so what music (or viewing history) should they have access to?

In truth smart speakers don't handle this especially well. At first they didn't do so at all,

Siri, play "Waterfall" by The Stone Roses.

This request will be answered either with a rendition of an iconic tune from the post-Grunge, pre-Britpop era track (announced by its title enunciated excruciatingly clearly) or, more likely, an extract of the same track. You'll need to be logged in as a user with a music account that has access to the track to hear it in full. For Apple users that is Apple Music; Spotify is supported by Amazon and Google devices.

SIGNING IN

Seems a lot easier on a screen than by voice, but that will likely improve over time.

with an Amazon Echo, for example, linking only to one account. When Google entered the market, it set the trend for allowing multiple users, which (in theory) makes the smart speaker a useful way to access different features, such as your own Google Calendar.

Using TV-based apps tends to be a little better – it isn't natural to pop to the settings page on the TV to switch users before settling in to watch something, but you can do it with, for example, the Netflix app on an Apple TV (though not with many of the video apps). And with video it is especially useful as you can retain your "where I got to" settings from one TV to another, or an iPad.

Combined family accounts solve a related issue, allowing you to connect one credit card but retain the ability to remotely approve (or not) your children's attempts to watch TV or download music or apps.

I definitely recommend getting a family subscription; these will allow you to approve purchases of movies; on Android it's called "family plan." Apple also do this, but you do need to be brand loyal!

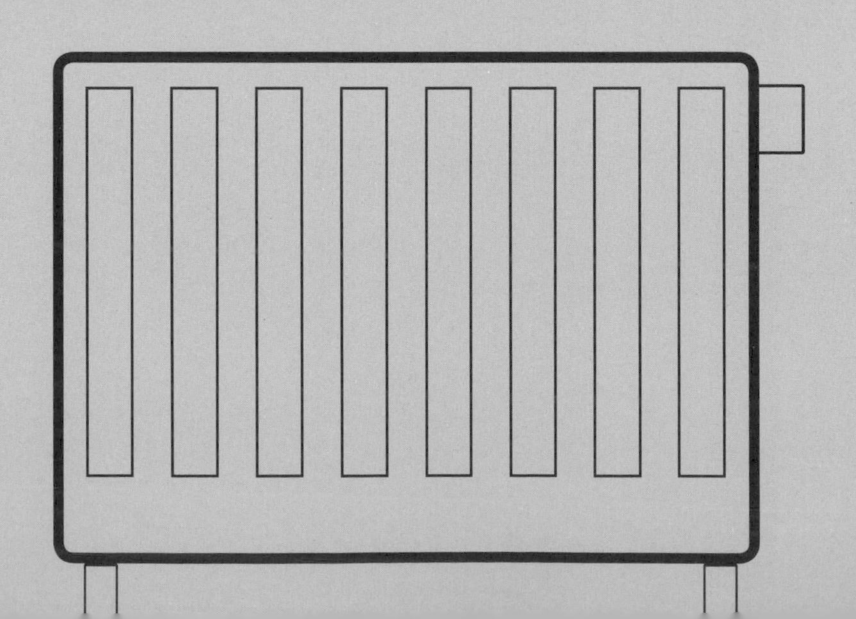

TEMPERATURE

One of the most practical applications of the Internet of Things is improving the way you control the boring home hardware — we're talking boilers and air-conditioning. If you get it right, you'll optimize your appliances' efficiency and save money.

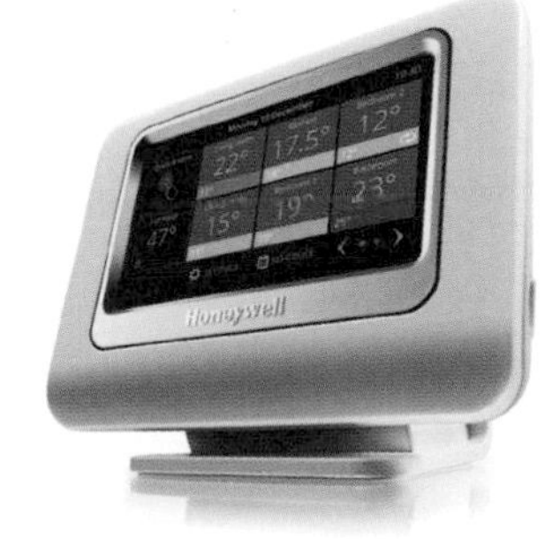

THERMOSTATS

Remote temperature control

More than a third of all energy use goes on heating and cooling homes, so the way you apply temperature control can have a huge effect on both global emissions and your own energy bills. Thermostats — temperature control switches — have been around for a long time: the first invented around 1620 by Dutch innovator Cornelis Drebbel, for incubating chickens).

Over time, Scottish chemist Andrew Ure's 1830s innovation of binding two different types of metal to each other and allowing the difference between their expansion to move the switch was adopted into the design. This kind of thermostat was the basis of the kind that has steadily made its way into homes for nearly 200 years. Standards have emerged — not the same everywhere, but enough that most central-heating systems can be controlled by some form of smart home thermostat.

As already noted, thermostats are also one of the first common home devices to be made smart. Unlike some of the other devices we'll encounter in these pages, the appeal and practicality of this improvement is obvious. Effective central-heating control is best achieved using a combination of a thermostat, which maintains a certain temperature as set by the user, when heating is needed, and a timer, which turns the system off altogether when heating is not required. It is far from unusual for one or another part of this equation to be missing even from a modern home, and where they do exist they're still often separate — the timer may be built into the boiler and the thermostat positioned on a wall, making for inconvenience when you want to tweak the settings.

Digital devices have steadily added flexibility, like the ability to set different temperatures at different times of day, and the ability to recognize and behave differently on weekends. Even with a device that can appreciate that not every day of the week is the same, however, digital controls still rely on you operating them finger-to-button, so to speak.

Step in the Internet of Things. In 2011, Nest introduced a device with some additional features that have gone on to define the expectations of a smart thermostat. Among the innovations they introduced was the ability for the thermostat to use your phone's GPS to determine whether you are in or near your home or not and disable the heating automatically if no one is home (helpfully, several users can be connected to the same thermostat). Another was its ability to build up a profile of its users' behavior over the weeks (the temperatures requested and at what time) and program itself.

Additionally, the device can take advantage of the internet to gather weather data so it has an idea how much work your boiler or AC is going to need to do to get the temperature where you want it, so if it's going to be particularly cold, for example, your boiler will get to work earlier to make sure you wake

EXTRA EFFICIENCY

Efficient control of your AC unit will save you money (if everyone did so, it might save the planet).

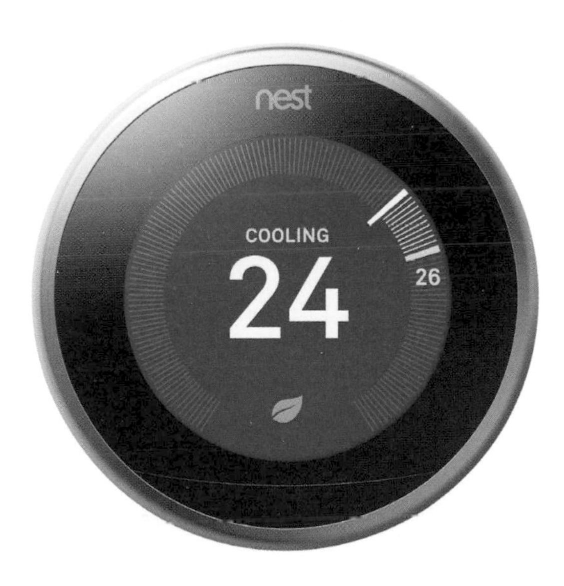

NEST

The Nest is an iconic smart thermostat, capable of being connected to both heating and AC systems. It builds up a user profile so that over time it can foresee your needs.

ENERGY SAVING TIPS THAT WORK – SMART OR NOT

- Don't turn the thermostat up higher than you want the temperature to get to; it won't make it get hot any faster.
- Don't turn it up higher when it's cold outside; the target temperature should be the same and the system will automatically do whatever extra work is needed to combat the cold.
- According to *The Telegraph*, 38 percent of people believe it's more energy efficient to leave the setting the same, constantly, but this is a fallacy. Turn it off when you're not home.

up or come home to a nice, toasty house. Because it works on Wi-Fi, you can move the thermostat itself to wherever it is needed (although it still needs to be plugged into a power supply), and an optional stand is sold by Nest. And, of course, while you can still program your temperature settings on the device itself, you can also do this on your phone via an app, which many find comes more naturally these days.

The practical upshot of this is that your thermostat is working hard to save you money and making your life easier in the process. The Nest thermostat, certainly, is very easy to use and is now offered in two versions (one capable of driving more kinds of boiler than the other).

Installation requirements depend somewhat on what you already have in your home. I have installed two Nests in my time and I found the second – in a new home with a new boiler and a very old thermostat – considerably harder, simply because my boiler came with extremely opaque documentation and there were no helpful tutorials for that model online. I would suggest that if you're comfortable shutting off power and you have good documentation for the boiler's connectors, then installing a Nest is not much harder than replacing a light switch. But for those who wouldn't feel comfortable doing that, or simply find their heating system inaccessibly located, call a professional – you may be able to include it as part of a regular service or upgrade.

Hey Siri, set the temperature to 78 degrees.

Sadly, in a home with a Nest-based thermostat, this had no effect. Although change has been hinted at by Nest – part of the Google group of companies – Apple's HomeKit is not currently supported by Nest.

Alexa, set the temperature to 21 degrees.

If this sounds a little uncomfortable, remember that your author is based in Europe. One of the nice things about digital systems is that you can switch between your preferred scale – Celsius or Fahrenheit. Once the Nest Skill has been added in the Alexa app, this command will work nicely.

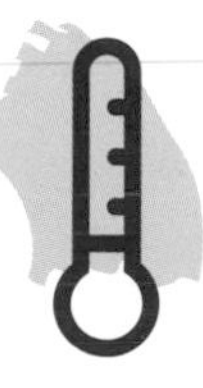

ECOBEE4

At a similar price point to the Nest 3G, the Ecobee4 offers two notable extra functions: an additional temperature sensor you can locate elsewhere to monitor consistency, and a built-in Amazon Alexa microphone. That doesn't just mean it can respond to orders from an Echo in the house – it also means you can ask the Ecobee4 to do almost anything (except play music) that an Echo can, saving you from filling the house with little smart speakers while extending voice control that bit further. Unlike Nest, it is also compatible with HomeKit.

I'm not actually sure our bills went down as we hoped; the device is great but it isn't as mean as I was with the old thermostat.

Honestly it seems the same to me. Newer looking. But the same as the old dial. Then again, we'd often leave that on the whole time because there was no timer.

ZONAL TEMPERATURE CONTROL

Adding additional thermostats

A single thermostat will not be ideal for a larger home or in a situation where you want some rooms to be kept cooler than others. Depending on the construction of the space you want to heat or cool, there are different options to consider.

At the grand end are zonal systems that include large boilers and cooling systems, with valves that direct heat or baffles that direct cool air to individual areas according to separate thermostats.

This kind of system, which is not dissimilar to that found in an office building, does potentially risk competing with itself. If one room set with a relatively cool temperature is right next to a neighboring space set to a higher one, you might find yourself paying for both sides of the heating and cooling equation, especially if doors get left open.

On the other hand, if used properly, you can make energy savings by setting less-used rooms to not be heated as often or as much, and, as a result, many users report that they achieve savings.

What about more modest properties, or situations where you're not interested in replacing your entire heating system? Well, if your heating is based on a radiator system, there is another solution that might be useful. Many governments have recently made much effort encouraging homeowners to fit thermostatic radiator valves that allow radiators to automatically turn themselves down when their own room is warm, rather than following the main thermostat's "one zone to rule them all" logic. Now smart versions of these valves are becoming available, operating on batteries. They certainly offer an appealing alternative and the disciplined user even more potential for savings.

SMART RADIATOR VALVE

This Netatmo smart radiator valve controls the individual radiator. It connects to an app with all the usual smart thermostat features, like the ability to automatically shut off when it detects an open window, and scheduling, so that you can set a heating routine for your entire house if you wish.

LARGE HOME

A very big property might benefit from multiple zones for heating.

AN EXAMPLE ZONAL SYSTEM

Here, a single boiler is connected to two heating zones using zone valves. Some smart thermostats can handle zoning standards, while others can't. Nest simply requires two complete installations, one for each valve.

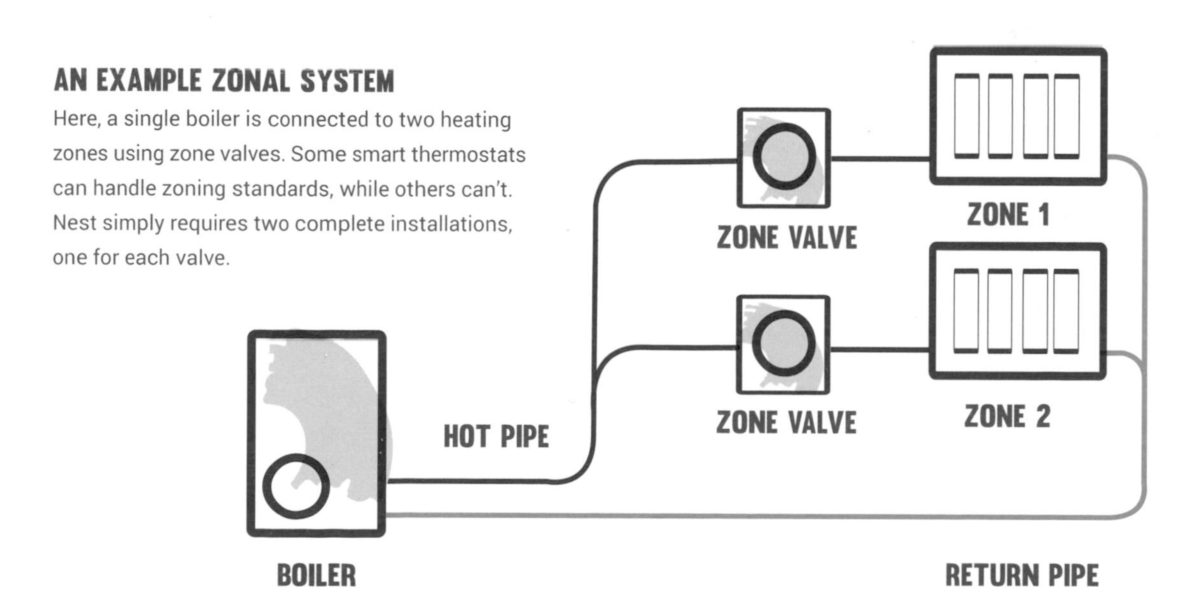

LIGHTING

One of the most visible areas of smart home technology is lighting; internet control has made both light switches and lightbulbs much more flexible — but more choice also creates some complications.

SMART LIGHTBULBS

Is it truly a lightbulb moment?

Smart lightbulbs have a great deal of appeal; the idea of being able to command your lights with a few words seems futuristic and exciting, and if we're honest there are plenty of times when that switch on the wall does seem a little far. Not only that, but we all know how to change a lightbulb.

It seems this is an achievable and appealing goal, so what are we waiting for, other than to work out what the best "how many smart home fans does it take to change a lightbulb?" joke is?

Actually, there are a few good reasons to pause before diving headfirst into the smart lighting trend. For one thing, if you are of the school of thought that says "if a job is worth doing, it's worth doing right," then you'll find yourself needing to replace every lightbulb in your home. I'll bet that if you stop to count them the number will seem quite high — especially when you bear in mind the price of a smart bulb (especially the cool, color-changing ones), which is much, much higher than for a normal bulb. In my fairly typical living room, the ceiling fitting has three bulbs, plus there is a standard light and two shelves with built-in lights.

In addition, a number of the lighting systems out there, including the well-established Hue from Philips, require a hub (which Philips calls a "bridge") of their own. This will be connected into your home network and will relay the commands from your phone's app or smart home assistant, as well as keeping the lightbulb's internal software up to date.

In exchange for these investments, you'll be able to turn your lights on and off using voice assistants, or from the other side of the world using your phone, as well as being able to adjust to movie mode (see page 73) on a whim without moving from your couch. That is pretty awesome. You can also add switching devices, like IR (Infra-Red) sensors, so that, for example, you can have an indoor light come on when someone steps onto your drive, which might give someone with nefarious intent rather more pause for thought than a traditional automatic security light.

Changing the color of your lights isn't just a gimmick, either. The ability to subtly shift the white point of the light, commonly known by photographers as the color temperature, allows you to take control of your moods and set light that will either help you work or help you rest.

In terms of compatibility, lighting is well supported by digital assistants, especially the established brands.

COLORED LIGHTS

The addition of color to bulbs, strips, and tiles allows you to customize your home decor to the *nth* degree.

PHILIPS HUE APP

The control screen simply allows you to touch an area to set the shade of light.

WHY YOU CAN ADJUST THE WHITE POINT

Traditional incandescent lightbulbs created a "warmer" light, and can be restful.

Pure white light is something we've become more used to from energy-saving LED lightbulbs.

A slightly "colder" light, such as you might get from a clear sky, is associated with concentration.

SENSIBLE LIGHT SETUP

Why naming your lights suddenly matters

"Dumb" lightbulbs don't need names. They come on when there is power and go off when there isn't; the switch is likely in the same room so you very quickly learn what switch turns off what light when you move to a new home – and you get some pretty definite visual feedback.

"Smart" lightbulbs are different. You can control them even when you can't see them, and you will do so either from a page in an app with potentially many others listed, or you'll be relying on a smart assistant which may well refuse to play ball without clearly named lights.

SMART LIGHTBULB SETUP

Remove the original lightbulb and place your new smart lightbulb into the socket. Manufacturers provide bulbs in plenty of standard fittings, so sadly the headache of working out which one yours needs isn't going away anytime soon!

Turn the bulb on. The bulb must be on when you get the hub up and running.

Connect the hub to the router using the ethernet cable then connect the power. Wait until it indicates connection.

Download the latest version of the manufacturer's app to your phone or Wi-Fi tablet and initiate the setup procedure.

NAMING TIPS

- It helps to think of your bulbs as being arranged hierarchically. Put the house/app name at the top, and individual bulbs at the bottom. In between, you can add groups. This is especially useful for fittings that take multiple bulbs.
- Name each bulb individually. If there is only one bulb, give it the room's name.
- The app might well have some pre-defined room names both as groups and as individual bulbs. It's best not to use both.
- Use short, sensible names or you'll regret it when using a voice command system!

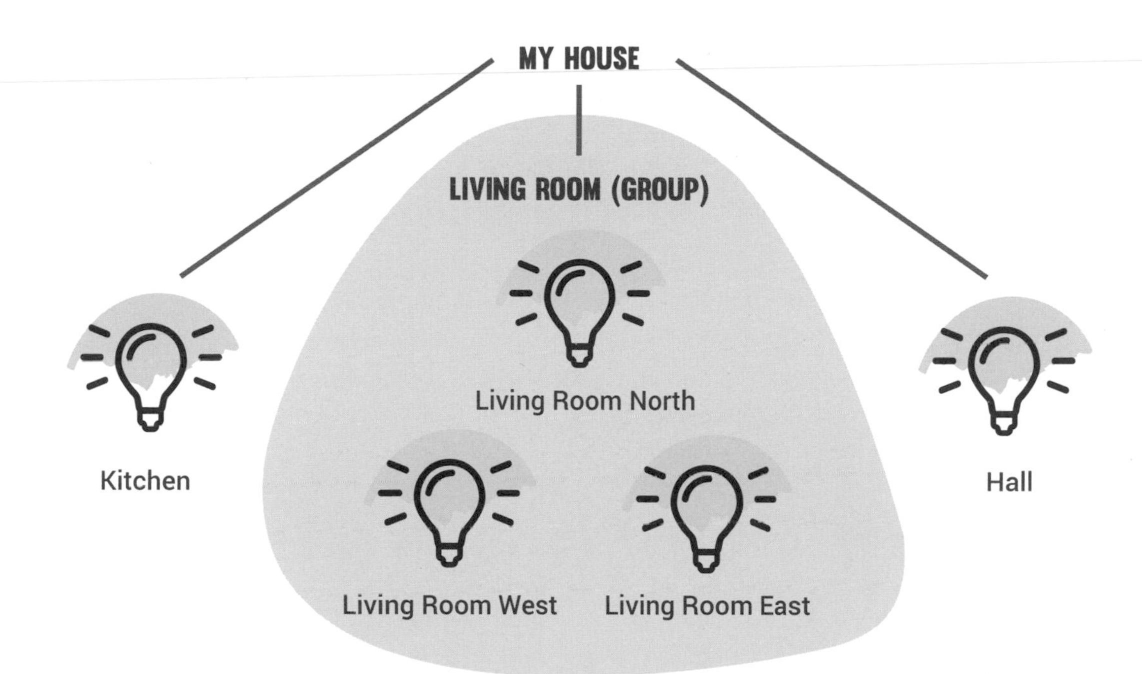

Hey Siri, turn off the living room lights.

"Got it," she says, and the "Living Room Lights" group goes off.

Hey Siri, turn on the lights in the living room.

"Sorry, I wasn't able to find anything like that this time." It turned out I got this error because my group had a different name.

Alexa, turn on the lights in the living room.

"I found more than one device with the name living room. Please give them unique names then run discovery again, or create a group if you want to control them simultaneously." This is a frustrating response as I have already created such a group in my Philips Hue app.

HUE VS THE OTHERS

With more brands joining, what is the best smart bulb?

The great fear for anyone buying smart home devices is the risk of owning "abandonware" — a product that is no longer serviced or maintained by its manufacturer. You need, after all, the developer of your chosen smart lights to stay trading and to keep their app up to date to fit the whims of your phone provider and so on.

Philips may seem like the safe option, but it does have competition from some well-established brands. Ikea's Trådfri range, for example, is accessibly priced, and offers adjustable white and dimmable bulbs — but its app is incredibly basic (you can't control the lights from outside your home network, and automation is minimal), although integrating it with HomeKit or Alexa improves your control.

LIFX already have a similar offering to Philips, with bright, color-changing bulbs and LED strips in the mix, and even some color-changing tiles to arrange on your wall, all controlled by a slick app and compatible with every major digital assistant. One nice touch is bulbs with a night-vision mode that invisibly provide the IR light your security camera needs.

One thing to be mindful of when choosing bulbs is the overall brightness. Color-changing bulbs often put out less light than you may need, so look out for the lumens rating before you buy. If your memory stretches back to the time of incandescent bulbs, you may remember how the energy burned was equated to the light given off: 40W would be quite dim while 100W was very bright. A bulb that puts out around 1,000 lumens will be broadly equivalent to the 100W, while 600 lumens is equivalent to about 50W. Check the specifications carefully.

IKEA TRÅDFRI

Stylish and affordable, but its limited app and lack of Google integration go to show that the biggest names don't always offer the best products.

LIFX

Next to Philips's effort, LIFX offers probably the most polished app, and has a great range of bulbs that are both colorful and bright.

HUE BULB

Even Hue, one of the biggest brands, offer only one maximum brightness option – equivalent to a 60W incandescent – for their colored lights.

LIGHTIFY & HUB

Cheaper than Philips's bulbs, Osram's Lightify bulbs do lack some of the fun features of Hue, but they have a good range of styles.

LUXJET BULB

Typical of the cheaper color-changing bulbs found via online retailers, this model offers a lower overall maximum brightness than the already-not-stunning offerings from established brands.

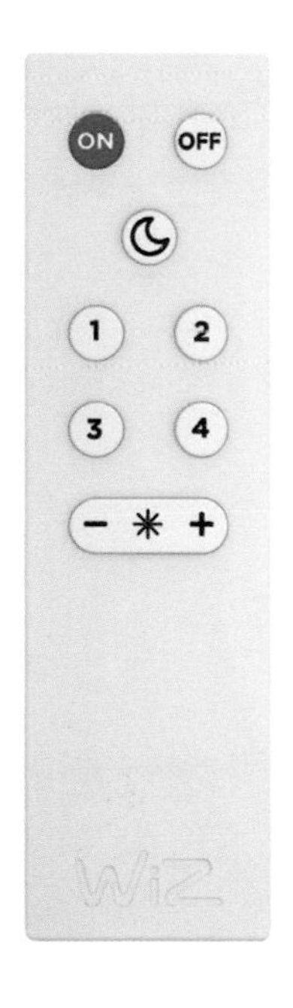

WIZ BULB

WiZ have an exciting range of bulbs and lights which don't require a hub (they use Wi-Fi) and are supplied with a remote. Integration with Alexa could be better.

SMART LIGHT APPS

Extending the possibilities

All the features built into your lights (dimming, color-changing, etc.) are generally accessible through an app provided by the manufacturer (and hopefully kept up to date during the bulb's lifetime). In addition, basic functions can often be controlled by voice assistants. But that's not the end of the story.

By opening up their code to developers, other companies have been able to design their own apps for Philips Hue, enabling the lights to do things the manufacturer itself had never dreamed of. A third-party app might, for example, simply provide a different way of selecting the bulb's color or brightness. But they might also do something altogether different, like allow your lights to respond to music or the TV, using your phone's sensors. This is nothing less (and nothing more) than a way to make your house more fun.

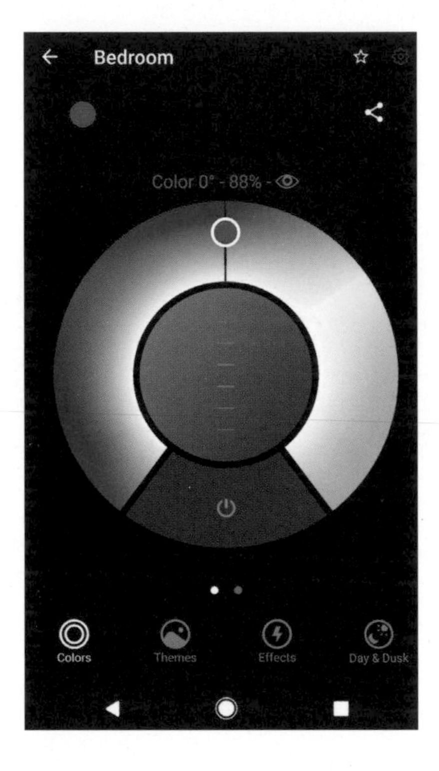

LIFX APP

The standard LIFX app will, as well as assisting with the setting up and naming of your bulbs, allow you to change the color with a tap with an easy-to-use color wheel, plus set up routines, choose preset "themes," and save your favorite configurations.

MIPOW PLAYBULB

MIPOW take things one step further than most by including an adequate, if not hi-fi, speaker in their bulb. Their app includes a mode that responds to the phone's motion so that – if you feel so inclined – you can entertain the kids by dancing to the music and the lighting will respond in sync.

TELEVISION BEYOND THE SCREEN

Philips offers colored stick-on light strips which you can stick onto the backside of a TV, using the wall to create a deflected ambient glow. Where it really gets exciting, though, is the extra features added by third-party developers. One of the most fun is Hue Camera, which works if you rest your phone somewhere where its camera has a clear view of the TV (a GorillaPod tripod for phones can prove useful), then crop the view on just the TV screen (as you would the camera zoom). Tell the app to control the Hue light(s) behind/near your TV and the surrounding glow will constantly change to reflect the main tone on the screen. The only slight issue with this is lag (the processing delay between the TV changing color and the light responding), but it's still awesome.

PHYSICAL SWITCHES

Overcoming the downside of smart lights

One thing that I guarantee you will realize very soon after installing your first smart lightbulb is just how often you'll still want to use the wall switch. As you reach for the old model, and experience the differences in the way the new one responds, you'll also begin to understand just what a simple, perfect system you've decided to change. And that's before we've considered the reactions of those you live with who maybe weren't as excited in the first place!

A smart bulb introduces confusion into a system we understand instinctively. Let's say you replace the living room light, which has a single switch, with a smart bulb. You switch the bulb on, and set it up. Then we turn it off with our phone and the light goes out. But it's still on at the wall switch.

What happens the next time you try to turn it on? Perhaps you casually flick the wall switch, as you're used to doing. But now you're actually turning it off. The light doesn't come on, and now you can't turn it on from your phone either. Turning it back on at the wall will now turn it on (possibly to maximum white, possibly to its previous setting), so this doesn't constitute a major problem, but it will probably give you a moment's pause. Essentially an issue that software developers call "usability" has entered your home.

You (and those you live with) might well decide that you'll simply live with this issue — you learn to just leave the switch on all the time — but your visitors might find things more confusing. This is where providing smart switches can solve things. If you have a switch that is just as easy to use, and part of the smart system, then you'll

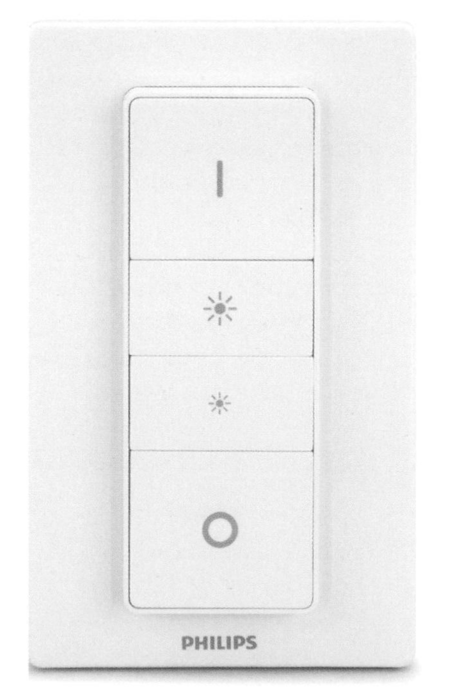

HUE SWITCH

This wall switch has a battery that lasts around two years, and four buttons: on, brighter, dimmer, and off. Other companies have similar offerings. The four switches actually form a remote that attaches to the surround magnetically (useful, but don't tell the kids — they'll hide it). You can use this switch instead of the wall switch to eliminate errors (all the control will be in the hands of the smart system), but mounting isn't easy. You may be able to stick it on top of the original switch, or put it next to it; the latter solution not really helping with the on/off uncertainty as well as being ugly.

TRUST SWITCH

Zigbee-certified, this can, theoretically, be set up to control different smart lighting systems.

TOO MANY SWITCHES

It's desirable to conceal the original switches to make using the new ones more natural, but not always easy! Then again, switches next to each other are both inelegant and confusing.

have all the "features" you're used to – a light that turns on and off at the wall switch – plus all the extra conveniences offered by the Internet of Things – the remote control.

Manufacturers have been quick on the uptake, and switches that are easy to install are not hard to find. They can be linked to the bulbs using the manufacturer's app. If you've got a corridor with a switch at either end, you'll want to replace both the old switches to eliminate the problem of the mains being turned off by a "real" switch. Double the price, but perfectly achievable.

A remaining issue is "latency" (the tech geeks' favorite word for the delay between operating a switch and the result). Radio-control enthusiasts and gamers obsess over the milliseconds of delay between command and the action, which can cause crashes.

The idea of a pause between switching your light on and illumination being as troubling seems unlikely; yet it is an issue. Signals to control your lights now often head via the hub and – in larger homes – through a whole mesh network, each introducing a minor delay. It's not disastrous, but it is a little weird.

Finally, for the sake of compatibility, it may be wisest to buy into an ecosystem (like Philips's Hue) that offers the full set of gadgets. No doubt you'll be able to find a switch that's in theory compatible with whatever bulb you choose, but getting them to recognize each other in real life may be a different matter.

SMART SWITCHES

Smart lighting, the other way

Changing your lightbulbs is probably the easiest and most obvious way to smarten your home's lighting controls – after all, anyone can change a lightbulb. But there is another way: replacing the wall switches themselves and leaving the "dumb" bulbs in place.

Smart bulbs offer the very visible appeal of color-changing fun. But it's very likely that the ability to turn your room purple at a moment's notice wasn't foremost on your wishlist before the bulb manufacturers began their advertising onslaught. Replacing the switch gives you access to some of the most useful aspects of smart lighting, such as being able to control lights from your phone, with your voice, or as part of a routine.

Beginning by turning your switch smart also will not prevent you from later installing a smart bulb – you just won't need to do so in places you don't feel the need.

Replacing the wall switches is definitely a little less straightforward than swapping a bulb. You're going to need to shut off the main power and be confident enough to take a screw-driver to your electrical wiring. In exchange, you'll end up with a more elegant solution (no extra switches on the wall) which is simple to operate.

The downside is that you might not have individual wall switches for all the lightbulbs you want to control (i.e., if they're wired as part of a bank, that's the way they'll stay). If you want that level of control, a partial solution would be mixing systems – smart bulbs when you have no choice, smart switches otherwise – but then you won't have control from a single app (though a single smart assistant might provide some consistency).

You also won't be able to control your plug-in lights through a wall switch, but you can solve this issue with a smart plug (see overleaf).

LIGHTWAVE SWITCH

This elegant switch system does require its own hub, but for European customers they'll find it neatly replaces standard light fittings, using exactly the same wiring. It's also fully compatible with Apple's HomeKit.

WARNING: PAIRING

When linking to Alexa, make sure the switch's app and the Echo are paired to the same account for voice control to work.

OITTM SMART SWITCH

Customers in the USA will find a number of generic replacements for the standard home light switch that will connect to the Wi-Fi (no hub) and offer apps and Skills for iOS, Android, Alexa, and Google Assistant.

BRILLIANT SWITCH

That's actually the product's name, but it's not an unreasonable assessment. For around the price of a tablet computer, this wall-mounted device has a screen allowing you to connect to an assortment of smart home tech, including a video camera so you can use a video doorbell. The touch-sensitive strips act as dimmers. There's also Amazon Alexa (mic and speaker) built in. At the time of writing, only 120v (USA) versions were available, however that's set to change.

RELAYS (SMART PLUGS)

Remote switches for table-lamps and more

For a very minimal outlay, it's long been possible to pick up a timer switch to use in your mains outlet. They're ideal for all sorts of things, like turning off decorative lights overnight. They can control the power of anything that plugs into an outlet, but are best suited to simple devices like lights; devices like televisions often also require you to press a button on a remote to turn them on.

Smart versions of these devices can be picked up either as one-off products that'll work with Wi-Fi and an app or those designed to fit into a smart home system using one hub or another.

As I noted above, be mindful of the simplicity of the classic on-and-off switch. One of the most common marketing lines given about these devices is the ability to turn on your coffee machine the moment you rise from your bed (or even before), but how sure are you that the simple arrival of mains current to the machine will do what you need? Do you need to flick any other switches? Does the machine have water, coffee, and a filter ready to go? What is the worst that can happen if it comes on without you being around to keep an eye on it?

All those caveats in mind, smart plugs are definitely useful. For one thing, the app collects all kinds of data, so you can monitor energy usage, and, of course, the ability to remotely power down the TV in the kids' room can give your parenting power an extra kick!

GE DIMMER SMART PLUG

Ideal for lighting, this smart plug can also control the voltage, allowing you to dim the light.

USA / EU / UK

You'll be able to find accessibly priced smart switches for all socket types.

LIGHTWAVE SOCKET

Just as you can with light switches, you'll also be able to find smart wall outlets for all the major systems.

I like being able to monitor the energy being used in my home, and using smart plugs lets me do this. I can also make absolutely sure the kids aren't watching their TV after bedtime.

KITCHEN

Moving on from voice-activated kitchen radios and speakers, internet technology has started looking elsewhere. Fridges, ovens, and hobs beware!

OVENS & HOBS

Now your oven can offer more than a timer

A timer is an incredibly useful addition to an oven and — slightly less so — a stovetop. Most modern ovens already feature a simple digital clock with the ability to turn the oven on or off at an arbitrary time, or simply sound an alarm. Smart tech offers something new: remote control.

As with other products here, the usefulness of that depends on how you use it (until your home robot comes along, you'll still need to physically put the food into the oven, and it isn't going to work very well for complicated recipes that require different stages, or lots of stirring, etc.). If, however, you possess the organization to prepare meals in advance, a smart oven affords you the flexibility to head out knowing you can return at a time of your choosing, rather than in time to meet a traditional timer. You can't add this function retroactively with a smart plug; ovens tend to be connected to other, more powerful mains supplies with their own fuses.

Appliance companies haven't stopped with remote controls. Whirlpool have added Yummly recipes to their hobs, ovens, and microwaves, and they can automatically preheat themselves to the temperatures suggested by the recipe you choose, will cook for the time suggested, and turn themselves off automatically. (They also have Amazon Alexa functionality built in, Google Assistant compatibility, and the ability to notify you when your food is ready via your Apple Watch.)

Over the coming years, image recognition and the ability to automatically determine the cooking settings are likely to become cleverer and more common features. Until then, the current smart features are handy, but you pay a lot of money for a small convenience.

AEG WI-FI CAMERA

Many smart ovens allow remote monitoring using a built-in camera via Wi-Fi.

WHIRLPOOL-YUMMLY INTEGRATION

This Whirlpool oven offers recipe access and control via voice and app. As well as talking to your oven, the Yummly app lets you curate your own digital recipe book, with millions of recipes to choose from the ability to create taste and nutrition profiles to refine your searches.

THE SMART FRIDGE

The fridge is a natural center of the home

Making the fridge part of the smart home has been an ambition of major electronics firms since one was first shown at the PC World electronics fair in Japan in 1998, with correspondent Michael Stroh saying, "they were a passport to the future where everyone would live like the Jetsons."

As it turns out, customers were somewhat reluctant to invest in the touchscreen fridge at first and have remained so for two decades. The reason would seem to be the balance between usefulness and cost. The cost has remained high – at least double that of a typical American-style twin-door refrigerator – but at least the functionality has started to show signs of improvement.

Part of that is the fact that, since 1998 especially, we have all allowed more of our lives to be recorded digitally. The iPhone wasn't launched until 2007, and, in the time since then, digital calendars have replaced diaries for many (but far from all). Many of us also actually use the notes feature in our portable device, and the idea of a live link between a shopping list on the fridge and our pockets is not a bad one. Better still, many new smart fridges boast cameras that allow us to view the inside of our fridge (yes, the light will come on automatically) from the supermarket. This can come in very handy.

The fridge has long been the natural heart of the home – the place we can keep our myriad assorted pieces of information, from to-do lists to calendars to family photographs. In theory this should translate well into the IoT, offering added functionality – for example, the ability to play music or order groceries direct from your fridge.

The issue (assuming cost isn't one), however, is complexity – at every level. We all know what we're doing with scraps of paper and some magnets, but even with an easy-to-use interface, we're still awkwardly tapping at keys on a screen or relying on voice recognition to create a shopping list. In terms of the information on display, using multiple calendar accounts introduces complexity. Or what if all our playlists are on Apple Music and your fridge isn't compatible?

The point is, getting the most out of a smart fridge requires a certain amount of effort and is limited by the software; on a traditional calendar, for example, its very easy to highlight more important things with a big red pen – digital calendars tend to make all events look broadly similar. On the other hand, you can end up with a much tidier kitchen by eliminating those calendars, scraps of paper, recipe books, and other screens.

If you have the patience, you can drag countdown timers onto the items in your fridge (on the screen or the app). This lets you track when food has gone off, and see this from anywhere, but if you're prepared to spend that long labeling it, you're probably organized enough to be on top of this already.

SAMSUNG SMART FRIDGE

This modern smart fridge doesn't look too unusual, and the software provides access to the kind of things we might stick to our fridges anyway; a calendar, a shopping list, a page we can write on with our fingers, and access to music-streaming services.

I must admit that I have some privacy concerns about this technology, and there's less space for our son's crayon masterpieces.

GROCERY ORDERING

Never go to the supermarket again

The idea of avoiding visits to the supermarket is incredibly appealing, and the smart home is offering a number of different solutions to this "problem" that take advantage of the growing number of delivery services.

If you've ever used a recipe book, you'll no doubt have found yourself experiencing some surprise at the "everyday ingredients" the author assumed you'd have. Now, theoretically, you can have your fridge make sure not only do you have these foods, but fresh too.

There are several different ways you can get groceries delivered online. The two leading ones are supermarket deliveries and box services. Supermarkets offer delivery services and access to a significant part of their range via a website or app. Box services, which tend to be subscription-based, decide for you what they'll send (based on your preferences) and provide you with all the ingredients to cook a specific meal/meals. This can offer interesting new meals and cooking experiences that might broaden your palate – or it can be a recipe for disaster, requiring you to pay for food it turns out you don't enjoy, until you can manage to cancel.

When you use a supermarket delivery, you will likely be asked to open an account, and the service may attempt to automatically sell you the same things the following week, or perhaps build up a more sophisticated way of monitoring your usage of different products. This seems very helpful at first, but there is a downside; requiring so little input, you're likely to end up getting the same recommended products over and over again, and life can get very boring that way. You're also less likely to spot offers, choosing the convenience of a one-click shop over browsing more slowly and thoughtfully. Research has shown that you'll save more by changing what you order and picking up the bargains; the same is even more true with delivery. The best approach is to switch supermarkets each time so that the systems think they need to win you back.

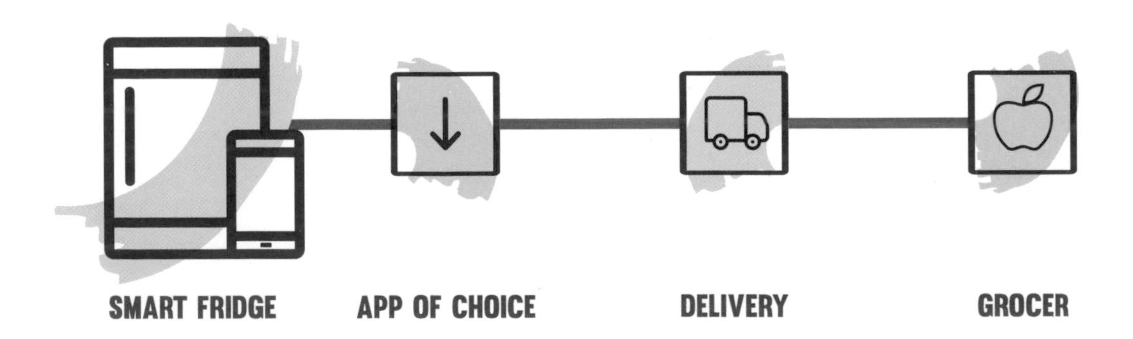

AMAZON DASH

Since it entered the grocery delivery fray, Amazon have decided to make buying as easy as possible; a cheap button linked to one of your essential products, which you can stick on to, say your dishwasher or your fridge (or keep in a drawer), and press to order more as soon as you see your supplies are running low.

WHISK APP & SMART FRIDGE

Here the Samsung Family Hub fridge is playing host to the Whisk app, a home for recipes that you can cook in your kitchen and reorder from participating supermarkets. The one shown is Whisk (UK) but similar platforms are available in different countries (including Instacart in the USA).

SMART BINS

Robot recycling coming soon?

It's no exaggeration (though perhaps a little sanctimonious) to say that sorting your bins has the potential to extend the existence of your species. Better recycling starts at home, and for those looking a little further down the smart home line, technology is nearing the market that is able to take it on.

The issue that many have yet to account for in their designs is limited space in the home. Sorting-and-compressing machinery adds to the already irksome volume of bins. For now the smartest thing to do with your garbage is simply to empty the bin a little more often.

BIN-E

Just starting to reach the market in 2018, a bin able to examine your waste with object-recognition cameras, before compressing and segregating it into one of four separate containers: mixed, glass, plastic, and paper. Prototypes have the ability to summon collection services once the bins are filled.

COMPARTMENT BIN

This elegant kitchen bin by simplehuman is typical of the modern age, with a recycling compartment and another for landfill. The local collectors are still depended upon for the mucky business of further sorting the recycling into its components.

KITCHEN DEVICES

Smartening up everything

It's possible to add remote control and an app to pretty much any device, and kitchenware is no exception. Since in general devices in the kitchen need outlet power, most of these devices will be perfectly happy with Wi-Fi, so there's no need to worry about hubs.

There is a genuine concern about longevity though; its more than likely that a product's app will become out of date as your phone operating system is updated, so don't let the remote functionality be the only influence in your choice of device.

At present, the devices range from the possibly quite useful (coffeemakers, kettles, and scales) to the probably quite unnecessary (such as iFAVINE's iSommelier, which offers improved aeration for your red wine).

KETTLES & COFFEEMAKERS

While you might wonder exactly how valuable is the power to "remote boil from anywhere," Smarter's iKettle is a very sleek device and comes with some nifty extra features like the ability to stay warm on "standby" mode. Like the iKettle, Smarter's coffee machine can easily be added to your morning routine, either by programming it to come on at a certain time as part of a routine with the Alexa, or triggered by something else with an IFTTT Applet (see page 130). Unfortunately for iOS customers, despite the kettle's distinctly Apple-esque name, these are currently only supported by Google and Amazon's assistants.

MONITORING & SECURITY

Security equipment like burglar alarms has long been able to keep an eye out for nefarious activity. Now those sensors can activate other devices, and your burglar alarm can reach you even when you're a long way away.

CAMERAS

Monitor, record, and be notified automatically

Video monitoring has gently disappeared into the background of our lives, employed, as it is so frequently, to be almost unnoticed. CCTV is more often than not a commercial or state security offering, but cameras are also found in most modern baby monitors, so the idea of using the internet to make cameras truly useful in more contexts is hardly surprising, despite the privacy issues.

This technology isn't perfect, and can conflict with other smart home staples. For example: I set my Nest to monitor my front room while away, and the lights to come on and off to simulate activity. Every time the light came on or went off, I'd receive a notification on my phone 5,000 km away. To its credit, it never said there was a person, just "activity," but it can still be unnerving.

In fact it's a lot easier to generate a list of good reasons for installing smart cameras than you might first imagine. You can use them at work to check older children have arrived safely home from school, to monitor the doorway in conjunction with a smart doorbell and smart lock (see pages 98–99) so you can receive deliveries from anywhere, or simply for checking in — if you're away on vacation, you can know for sure your house hasn't been broken into rather than just hoping so.

Internet-enabled cameras offer more than just online remote monitoring too, most are capable of monitoring the scene using artificial intelligence and letting you know (via your phone's notification system) whether they've seen something.

NEST CAMERA

Designed for indoor use, this camera works in normal light or in the dark using built-in night lights (hidden in the shiny area around the lens). The Nest camera can identify a significant change and/or the presence of a person, and (optionally) send a notification accordingly.

SAMSUNG WI-FI IP SMARTCAM 1080P HD PRO MOTION

This camera features wide dynamic range to allow it to handle bright backgrounds and dimly lit foregrounds, giving you a clear view of what it's seeing.

FEATURES TO LOOK FOR:

- **Monitoring/Recording** – What Nest call "Nest Aware" is a monthly subscription service that monitors the video output, looks for discernible activity using AI (for example, spotting people), and remotely records snippets for a period of time. A cynic might view this as a way of eking even more money out of an already expensive device.
- **Resolution** – Image quality is (partly) determined by the maximum resolution: 720p means 720 lines of picture; 1080p might not sound a lot more, but is actually twice as many pixels. You might find that is rather taxing on your internet connection though, so check that the quality is adjustable in the settings function of the associated app.
- **Two-way talk** – This featre means that you can use the app and your phone to talk remotely to someone in the room via a built-in speaker (you'll be able to see them, too).
- **Motion zone/Zone select** – You can limit the area of the video picture that your monitoring signal will track, so that, for example, an alert is only triggered if someone goes into a forbidden area.
- **Angle of view** – The widest angle is the biggest number, but can lead to a distorted, sphere-like image.

CAMERAS

Indoor cameras are a great way to watch pets; they can usually be powered using a USB socket and require relatively little power. Outdoor cameras tend to be a little more troublesome to connect up, so if you don't want to run power outside then look for (and remember to charge) a camera with batteries.

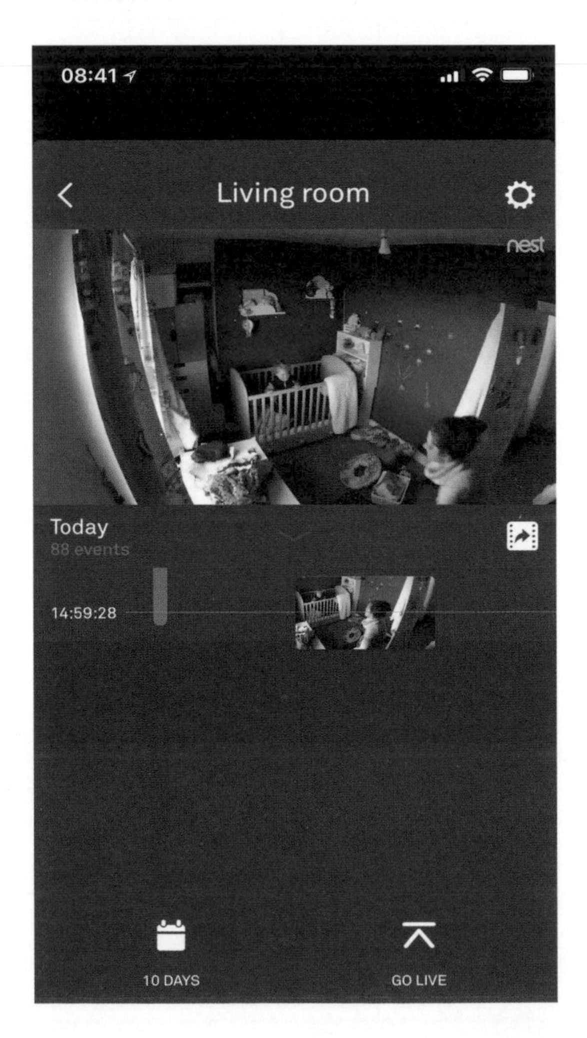

CANARY FLEX

This weatherproof camera can be either plugged in or operated using a rechargeable battery. Video storage requires paid membership.

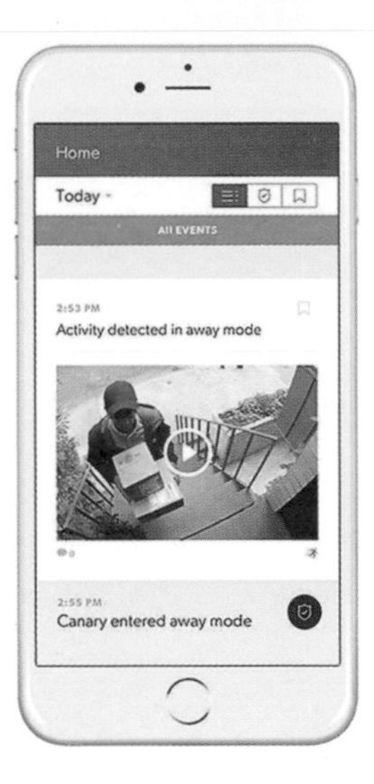

NEST APP

The Nest Cam app features the live view from the camera itself, and an area beneath it which provides clips recorded when the system viewed activity, assuming that you have chosen to subscribe to Nest's not inconsiderably priced monthly AI service.

RING FLOODLIGHT CAM

This weatherproof camera has a 140-degree field of view and all the usual features, plus (as well as a two-way speaker) a remote-operated siren that can be used to attract a bit more attention if you're confident your "guest" is unwelcome.

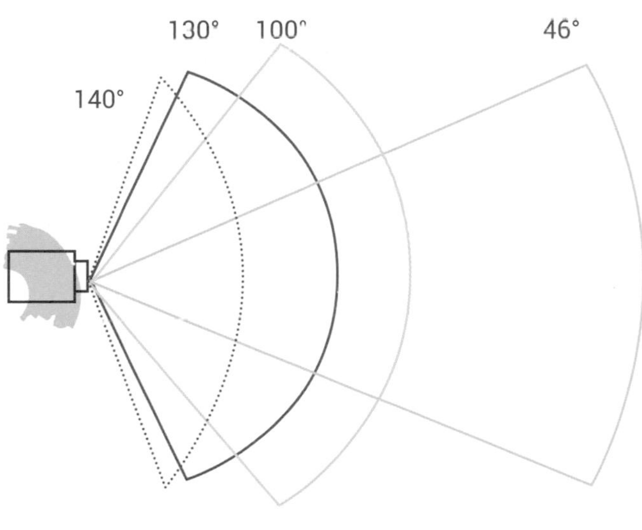

FIELD OF VIEW

This diagram illustrates the field of view the camera sees; i.e., what you'll see in the shot. Of course the image you're given is rectangular while this math requires a circular view, so the widest angles will give you some distortion. Photographers familiar with lens measurements might find the EFL (Equivalent Focal Length) comparison useful.

SENSORS

Motion sensors and switching

A smart camera with a subscription isn't the only way you can detect motion and act upon it. You're probably familiar with security lights that use IR sensors to turn themselves on when presence is detected. The smart version works in the same way, except that some can send you notifications when they're triggered, or can be configured to activate other devices via a service like IFTTT (see page 130).

This allows you to build up a much more sophisticated and flexible home security system, but, if you choose, you can still put motion sensors and door-opening detectors at the heart of it, and it's also possible to add an alarm siren.

The great thing for those with a more optimistic outlook on life is that this information is also available for positive uses, like turning down the thermostat when a window is open to avoid wasting money and energy on unnecessary heating.

I was excited to build a smart home security system, but compared to a traditional one, I found it was really expensive. Each sensor costs about four to six times what the standard kind does, and that starts to mount up if you have one for every window and door!

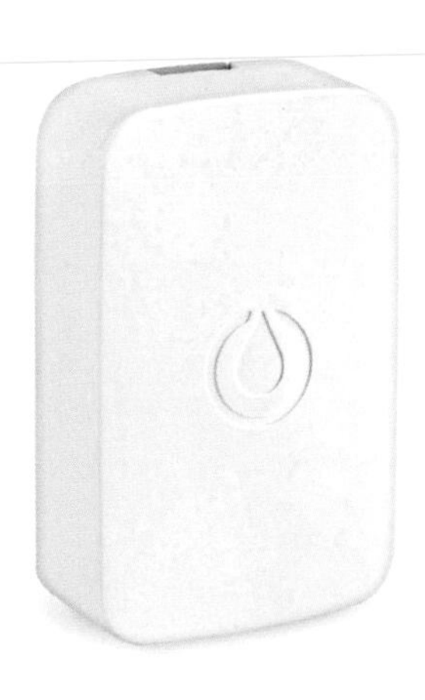

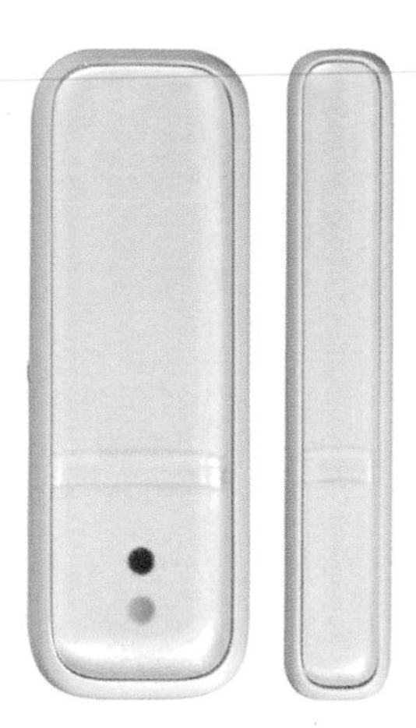

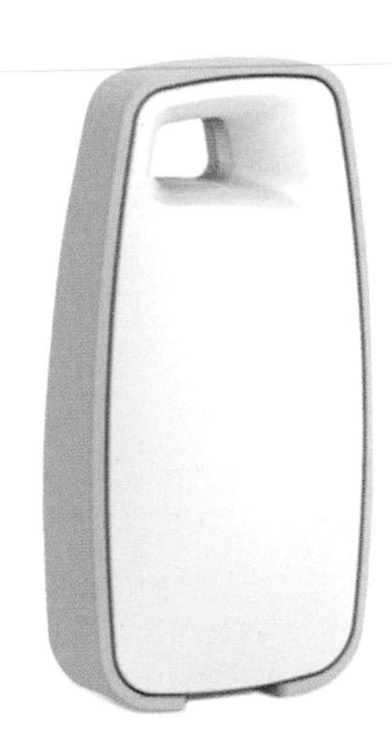

MOISTURE SENSOR

This sensor can be positioned in areas where there is a flood risk, like behind a washing machine or near the sink, so that should moisture start to gather you can be alerted before too much damage is caused (rather than when you get back from a trip!).

DOOR SENSOR

These use a magnetic "reed switch" to detect whether a door or window is open and are a long-standing feature on burglar alarms. Modern ones may also feature motion detectors.

PRESENCE SENSOR

A tool which you can use as a key fob that will help your Samsung SmartThings Hub determine whether you are nearby. This can be set, for example, to unlock the doors or turn on lights.

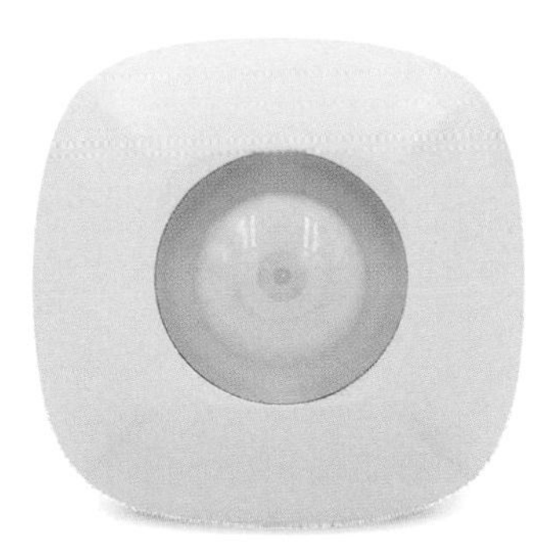

MOTION SENSOR

Using an infra-red sensor this device can detect movement from body heat. Some feature adjustable sensitivity and may also have a visible-light sensor built in too.

SMOKE/CARBON MONOXIDE SENSOR

Smoke detectors and dangerous-gas detectors can warn you of dangers other than intruders.

DOORBELLS & LOCKS

You've always got one eye at home

How often have you wished you'd been able to speak to the delivery agent when they visited and you weren't home? Or perhaps getting up to go to the door every time someone calls is something of a chore for you; and perhaps not every visitor is welcome. The connected doorbell takes one of the most mundane things in the house and gives it a potential that's easy to understand: a direct link to your phone.

The process of using the doorbell, once installed, is straightforward, though without adding the optional extra in-house chime, you're very dependent on your phone and an internet connection for something that should be more simple.

Locks not working is even more serious than a doorbell (be careful when ordering online – door types vary), and, privately, one of the representatives of a major company suggested to your author that this technology might be more trouble than it was worth.

SMART LOCK

This smart lock (shown from the front; the back for the indoor side has a battery compartment) can be opened by an app command, using a little chip you can put on your key fob, or using Bluetooth.

> **We thought the doorbell was a great idea, but then my mum came to visit and arrived before we could get back from work and, without a remotely controlled lock, I couldn't do much more than apologize remotely. A phone could do that.**

1

As someone approaches the doorbell the camera will see them. Some smart doorbells notify you at this point, or can be set to trigger a warning or a light.

RING

A connected doorbell with an optional traditional chime.

2A

Once the button is pressed, the doorbell will notify you via your app, assuming it (and you) has a good connection.

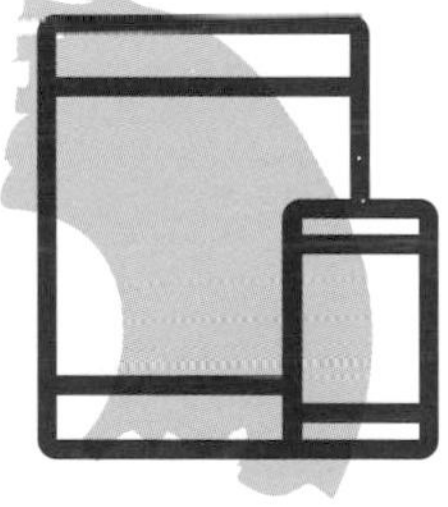

2B

If you have a chime in the house (often an optional extra), it will ring.

3A

You can choose to answer the doorbell via the app on your phone or device and talk to your visitor.

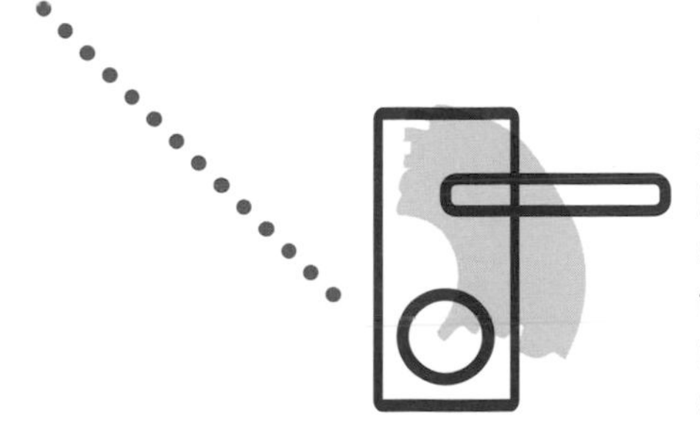

3B

If you have the ability, you might choose to unlock your door remotely too. The power to do this might well require another app, hub, and, of course, the lock itself.

BABY MONITORING

Wise old souls will tell you that in their day, simply putting the baby down to sleep and leaving them in their cot until the next morning was enough, and the modern concept of constant monitoring is "soft." However, monitors have been around since 1937 with Isamu Noguchi's "radio nurse" and are one of the most widely used pieces of technology in homes with babies.

Smart technology adds to the traditional microphone-based monitors the same cameras that it offers for in-home security. These cameras often match the higher-end options in terms of specification, using infra-red lighting to allow them to see in the dark. They still use microphones, but the ability to also see if everything is.OK without having to traipse to the baby's room and risk waking him or her is definitely useful. You also gain the ability to press a button and speak back to the infant, a feature usually known as two-way talk (quite what the baby makes of this is another matter).

Price-wise, what you get for your investment is mixed; baby monitoring is a competitive market, but some vendors of high-end internet-enabled monitoring systems charge surprisingly large amounts while still expecting you to use your phone as the monitoring screen (with all the attendant risks that has — running out of battery, being distracted by a call, and so on). Others do include a monitoring device, just like a "dumb" monitor has to do.

It's worth noting that all generic smart cameras on the market list child monitoring as a key *raison d'etre*, and they do offer good features, including the ability to use your phone's notification system should they feel your attention is warranted, not to mention working beyond the limits of your home's network, so that you remain connected to your child wherever you are. On the other hand, they are less likely to include the features like light projection and soothing music that you can expect on high-end dedicated baby monitors or on app-controlled devices like the Dream Machine (pictured). However, if you are set on going for a system that can be picked up at long range from your home, perhaps you should also consider if you should really be outside the home while your baby is in it?

Ultimately, you might take the view that these devices are gimmicks — I have a non-smart baby monitor equipped with all the other cool features discussed, but I have to admit to nearly never using them. In which case, a smart camera that will have use beyond the early years starts to seem more practical; once your little one reaches a certain age you can repurpose the camera as a home security device.

TURNING HEADS

You'll find it at the pricier end of the spectrum, and no, it doesn't come with a monitor, but this Netgear ABC1000 certainly looks cute. It can remotely move its camera to change the angle of view (making it look more as if the "rabbit" is looking at its feet).

DREAM MACHINE

This Motorola audio monitor isn't too pricey and offers the ability to remotely turn on and off soothing projections and music, in addition to the usual monitoring service.

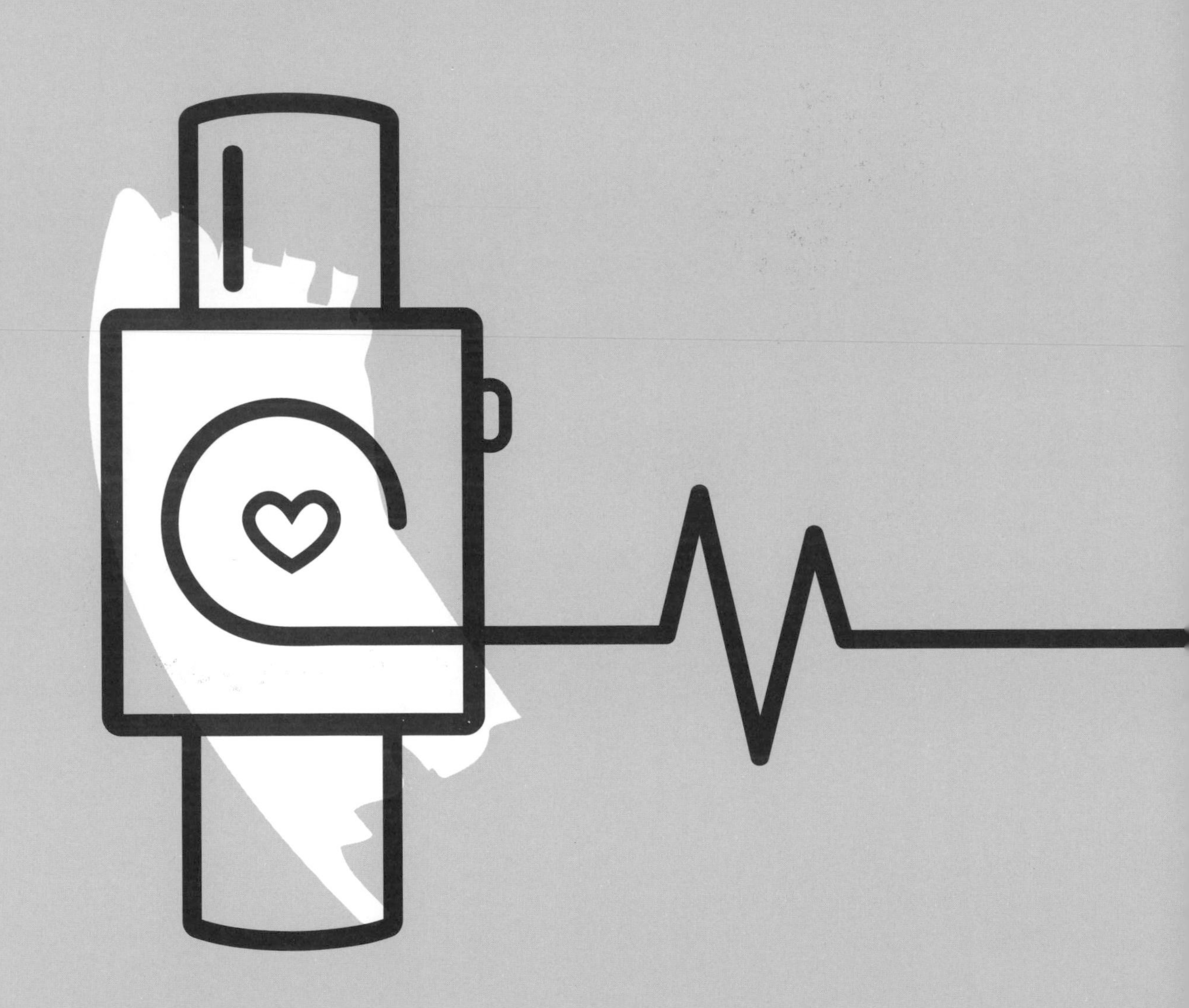

WELL-BEING

Most of us like to keep an eye on our health (although, if we're honest, we'd probably all prefer it to be a bit easier to do so). Technology won't make you healthier, but it does make it easier to set yourself goals — and stick to them.

WELL-BEING CONCEPTS

Machines to assess your lifestyle

A word like "well-being," which encompasses health and happiness, invites a number of different types of measurement. Some of these are easier for technology to portray than others, and it is beyond the scope of this book to decide whether they are actually the best measures, so here we will focus on those aspects the device manufacturers themselves have addressed, and which seem to be the most fashionable and desirable at present.

Perhaps the most established of physical characteristics that we measure at home — and one most of us haven't needed the prompt of digital technology for — is weight. With consideration of factors like height and other measurements, weight is not perfectly correlated with health, but it's a good start. Measurements, especially of the waist, are useful for building the full picture; but while a tape measure is distinctly manual, digital scales can offer data to an ongoing record.

With the arrival of devices like the Fitbit, plus smartwatches (see page 108 for more on smartwatches), a new market segment has been created, now with around 40 million devices, half of which are Apple's. With it, people have acquired the ability to track their daily movements to build a picture of their health and health/exercise/diet-related habits. The technology is amazing, though sometimes I'm rather inclined to think of my wristwatch as an angry coach as I prepare to let my country down in a sprint.

The tech is not perfect; if you want to track the calories you ingest, for example, you still need to keep a food diary — you may find an app to do it, but it's still relatively laborious compared to the exercise monitoring that is achieved by detecting your motion (and combining it with GPS data, plus, ideally, your instruction to the watch about what activity you're doing).

If this all feels like adding stress to an already hectic lifestyle, you won't be surprised to learn that the smart home world is starting to look at things from the other end of the scale. How can your level of rest be measured? Manufacturers, including Nokia, have produced pressure-pad devices that can be placed under the mattress to monitor the extent to which you toss and turn to give you

FITBIT
From 2015 it was practical to monitor your heart rate 24 hours a day with this Fitbit Charge.

WATCHES

Although Apple's is by far the most popular smartwatch on the market, there is now a far greater range of designs to choose from, including this one from Kate Spade, and Nokia's stylish Steel HR.

SMARTER SCALES

Scales don't just measure your weight anymore. Usually coming with an app, smart scales like this budget one from Chinese brand YUNMAI can give you information on bone mass, muscle, fat, and more.

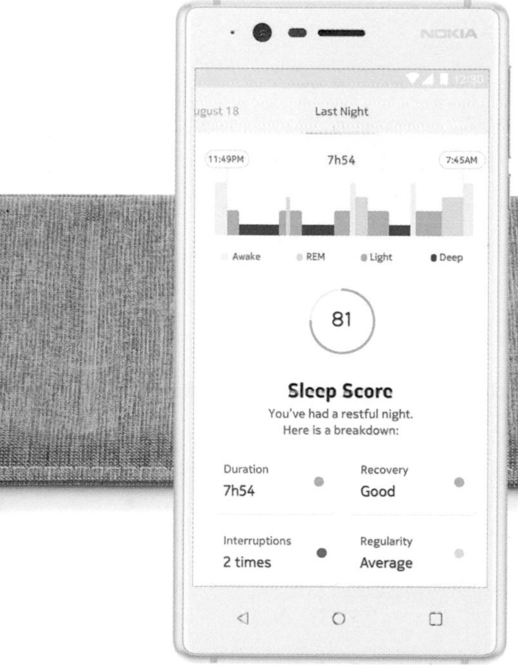

SLEEP SCORE

The Nokia Sleep Mate and accompanying app giving a sleep score readout.

a sleep score. A cynic might inquire as to how they're meant to change their sleep score, given that they're unlikely to be conscious during the critical period; but the data does have uses beyond mere monitoring. Correctly connected, for example, your alarm can be configured to wake you during a sleep cycle that will cause less grogginess. The accompanying app will also offer constructive behavior-based suggestions, called the "Sleep Smarter" program, to improve your sleep.

In addition, Nokia's sleep monitors can be set up through IFTTT to, for example, turn on your heating, or your coffee machine when you wake up (see page 128 for more on using IFTTT). If there are two of you sleeping in the same bed and you both want a score, you'll need a monitor for each side of the bed, but you will each get individual results so that you can set your own event triggers.

Monitoring the conditions of your bedroom as a whole is another growing area, with devices like the Netatmo (pictured opposite) addressing its air quality — a factor that may be significant in your sleep cycle and general health.

As well as these somewhat peripheral concerns, factors central to health, like exercise, often benefit from the kind of monitoring and diary-keeping that can even lead to personalized care suggestions (and a reduction in visits to the doctor). An early example of where this approach is already in use is the app PhysioTrack, which allows physiotherapy patients to track their progress, and which then prescribes exercises for the patients without the need to visit a practitioner. (Physio is a good area for this kind of technology to be trialled because it is a kind of therapy that isn't biased toward older patients, who may not be *au fait* with apps.)

LOOKING INTO THE FUTURE

The population of the UK is a little over 65 million (think California and Texas put together). The UK government is expecting the country to be home to 32 million people over the age of 60 by 2039, and similar levels of population growth and aging can be seen throughout the developed world. Why single out the UK? Because the government has started to look at how smart home technology can be used to combat traditional issues relating to an aging population, including housing needs. The government's white paper (discussion document) on the issue makes much of forming an "Agile Aging Alliance."

PHYSIOTRACK

You might know it by a different name, as PhysioTrack can be branded by a practitioner that licenses it from the developer, but if you're a patient you'll get personalized exercises without needing to visit your doctor or therapist.

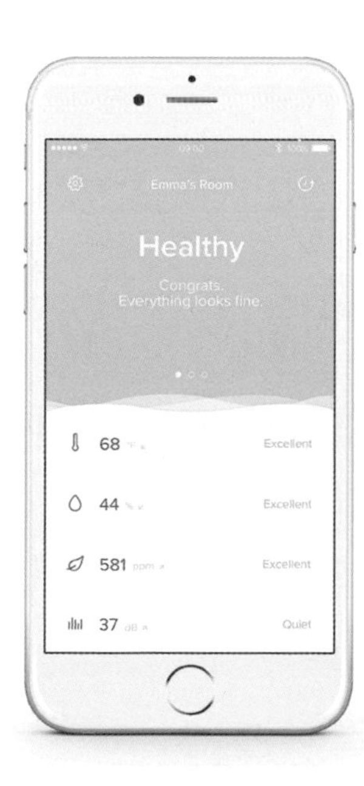

ROOM MONITOR

The Netatmo Healthy Home Coach monitors sound, air temperature, and air quality (humidity and pollution). This helps you flag any problems and correct them early on.

WATCH MEETS PHONE

Machines to assess your lifestyle

Activity monitors, whether built into a digital watch alongside a host of smart features or as a single-purpose fitness tracker (plus watch), have become a major part of the tech scene since the Fitbit first emerged in 2008.

One question that should be answered straight out of the gate, though, is whether these devices help you lose weight. The answer to that, at least according to a randomized study conducted in 2016 by the medical journal JAMA, is a fairly blunt "no." Simply having the data from a device did not enable overweight users to achieve better results than those without the monitors.

That's not to say that people trying to lose weight is the only target market; if anything, people who have always concentrated on their fitness may be more concerned with reaching their peak, and they will likely be at least as likely to benefit from the data offered by activity monitors.

Whatever your reason for monitoring your health, accuracy is the most important factor. How can a device that you keep on your wrist tell you how much exercise you're getting? The answer is that it makes something of an educated guess, and its ability to do so depends on what sensors it has on board or has access to.

One of the things commonly measured is heart rate. All Apple Watches, for example, monitor this using photoplethysmography, meaning that it flashes a green light underneath the watch very quickly and measures how much light returns; this waxes and wanes between each pulse, even at your wrist, providing enough data to work out your heart rate. Your heart rate is very useful in assessing how much exercise you're experiencing, since your heart tends to beat faster when you're working harder and burning more calories.

The popular Apple Watch also has a number of different settings for different types of workout, and each of these translates the vibration and movement that the watch senses into a calorie score. If the watch is able to track motion using GPS it can also work out how fast you are moving, which will be all the more helpful in getting an accurate running or cycling score. Apple specifically tell you to bring your iPhone along with you for the first 20 minutes of your exercise session if you're using a first generation watch, as, unlike the later versions, this did not have GPS built in. After this sample session, your watch will use the improved data from your phone to provide a more accurate score.

APPLE WATCH

The most popular smartwatch on the market, the newer versions include GPS for more accurate tracking.

FITBIT

Known as an activity tracker rather than a smartwatch, from a data-gathering perspective there is little in it, while its price is far more accessible than the smartwatch.

WORKOUT TYPE

This watch is being set to Outdoor Run with no specific goal, but you could choose a time or even calorie goal and be alerted when you reach it.

BODY FAT

Feature-rich scales

At the beginning of this chapter, when we looked at what could be measured, we noted that your weight alone isn't necessarily a good indication of your health. Your body fat is a much better measure, because it allows for the proportion of bone, muscle, ligaments, tendons, and organs that compose your body (all adding to your "good" weight) to be taken into account, and so it is a much better indicator of conditions related to obesity.

There are plenty of ways to measure this, too, from skin calipers (not especially easy or accurate) to extremely accurate DXA (Dual X-Ray Absorptiometry) tests that a hospital might conduct. Coming in somewhere between the two is the bioelectrical impedance method; also not perfectly accurate, but very straightforward for designers to include in their devices. It works by passing a current (don't worry, not a high one) through the body, for example, from one foot to another, on a set of scales.

On a leaner body, the current will travel faster, as fat impedes its progress, and the result is derived by measuring the speed at which the pulses of electricity travel. To get the best possible results using this kind of measurement (and with weight), you should aim to conduct the test at a similar point in each day, so you are similarly hydrated (water levels also affect the speed at which the current flows, thus affecting the score).

A good feature to look out for is the ability to remember individual users; some scales now cater to up to 16 users, which I've seen described as "perfect for sports teams or even small businesses planning wellness initiatives." The latter I'll leave you to keep your own counsel on, though if you are thinking of motivating your colleagues, I can imagine that getting the scales out might not get a 100-percent-positive response.

NUTRITION

Although well outside this book's frame of reference, monitoring the percentage of your body that is fat is the other end of the process from nutrition – looking at the balance of ingredients going in in the first place. Computer programmers use the acronym GIGO, which stands for "Garbage In, Garbage Out," and while it's definitely meant differently, assessing fat at the scales end is not the most effective way to bring about change. In the smart home, nutritional information will often be included in recipes which you can search for on devices like the Amazon Echo Show, and entered into an app for even more accurate health analysis.

WEIGHT VS FAT

Today's scales can tell you what percentage of your body mass is fat, as well as your weight. The table below gives you approximate categories (segments do vary with age).

	MEN	WOMEN
Dangerously low	Below 2%	Below 10%
Essential fat	2–5%	10–13%
Athlete-level fitness	6–13%	14–20%
Healthy	14–17%	21–24%
Normal to high	18–24%	25–31%
Obese	Above 25%	Above 32%

BLOOD PRESSURE

Data that really helps your doctor

Much of what can be measured at home is useful for tracking your exercise levels, but when you visit the doctor you'll find that they will have a few more questions. One of these will almost certainly be about your blood pressure.

Blood pressure (BP) is normally expressed as one number over another, which makes it sound a little like a math class. The first of these is the systolic pressure, which is the highest pressure reached during one beat. The second, or lower, is the lowest pressure reached between beats, and is called the diastolic pressure.

Both of these are measured in mmHg, referring to the millimeters of mercury moved in a traditional (but now unusual) measuring device, and the baseline is the surrounding atmospheric pressure. Normal resting blood pressure in an adult is 120/80 mmHg, and you're likely familiar with the semi-automatic cuffs that are commonly used by doctors, which have the advantage of not requiring any dangerous mercury in their construction.

It is important to monitor your blood pressure because it can reveal health issues early on. High blood pressure suggests that the heart is having to work too hard, probably because the arteries are becoming clogged. Persistent high pressure (known as hypertension), even at a fairly low level above the normal, is a clear risk factor in heart attacks, strokes, and aneurysms.

At the other end of the scale, low blood pressure (hypotension) can be a concern. Below a certain level, there may not be enough flow to get the blood around the body, so fainting is possible, or even shock (in the medical sense). Low blood pressure can also be caused by toxins, certain eating disorders, sepsis, or a hemorrhage (though, in all fairness to self-monitoring, if you're actually losing blood, dealing with this will be more important than getting an immediate measurement of blood pressure).

The idea of taking your own measurements to monitor health is well-founded, though; there is even evidence that measurements taken at night (when your doctor is less likely to be hanging around) are better indicators of your health than those taken in the day. That makes sense, not only because there is a 24-hour cycle to your pressure (higher in the morning and evening, lowest at night), but because during the day we're always doing something a little different. Perhaps you just had to rush across town to the doctor's before you had your pressure measured? What if at the previous appointment you'd had plenty of time to relax in the waiting room?

Qardio and Nokia are amongst the manufacturers who offer smart BP monitors, and in both cases they also integrate the data collection with other devices in their families (like scales), so you can accumulate data in one place — as long as you don't mind picking up a new set of scales at the same time as your BP monitor.

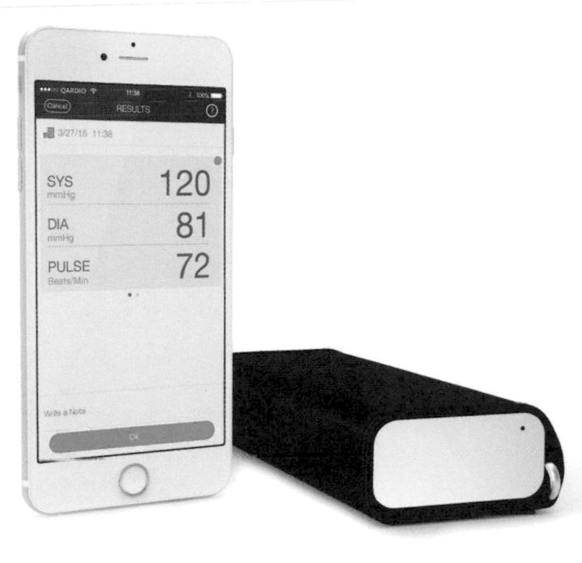

QARDIO SMART MONITOR

A portable blood-pressure monitor that provides BP and pulse measurements, and can email data directly to your doctor.

I didn't see the point of a smart monitor, I just kept a notebook next to a very cheap digital monitor, but then I went to a doctor's appointment without my diary of blood-pressure readings. If I'd had the "smart" one, the app would have been with me, and all those readings.

BLOOD-PRESSURE MONITOR

This "non-smart" electronic blood-pressure monitor can store up to 90 measurements in its memory, and will also take a pulse reading. It's less than a quarter of the cost of a "smart" one.

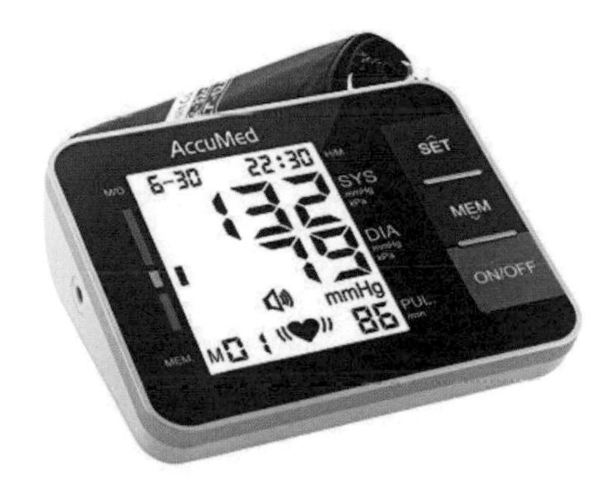

TEMPERATURE

No more mercury

Measuring body temperature has certainly become considerably safer in recent years, and rather more dignified, assuming you've chosen to keep up with the technology.

The old mercury-based thermometers had to be used with extreme caution and couldn't really be given to babies, let alone toddlers. They were replaced with versions using less dangerous chemicals, but still surrounded by glass. In my youth, it was not unusual for a liquid crystal strip to be held against the forehead and for the different squares to reveal themselves depending on the temperature. (Should you want a day home from school, these were pretty easy to trick with some fervent forehead rubbing.)

Now, digital thermometers are commonplace, some being capable of taking readings from the mouth or by being held in the armpit (the former being practically impossible and the latter being simply very difficult with a young toddler). As with other health-monitoring devices, the digital thermometer provides accurate readings. Adding smart home technology is mostly about tracking changes over time and displaying them to you in a clear and easy-to-understand way.

One brand, Kinsa, has developed an app that allows you to track the temperatures of multiple users. The probe connects to your phone's headphone socket (or, for iPhone users, the headphone adapter dongle), and can be set to Oral, Underarm, Rectal, or Ear mode. Best of all, while the temperature is being measured – always an awkward moment – a game pops up on the screen to keep you distracted.

This level of thoughtfulness in app development should be your main criteria when selecting a smart thermometer; those that seek to inform you what their reading means (fever, seriously ill, etc.) and what your next steps should be are the best examples of how smart tech can offer more.

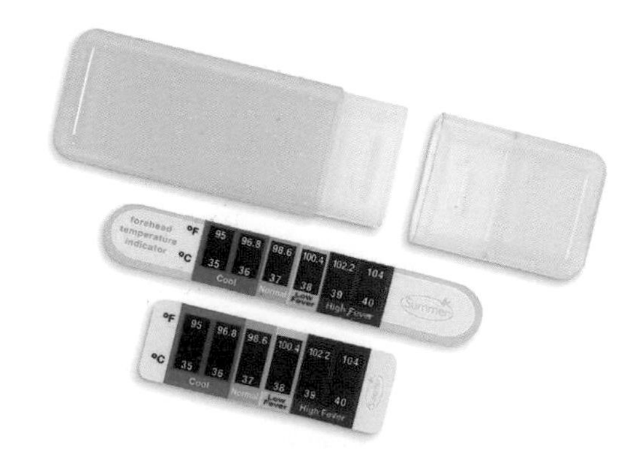

FOREHEAD TEMPERATURE INDICATOR

Gives a reading in around 15 seconds; a strip like this is inexpensive and useful for keeping in an emergency kit.

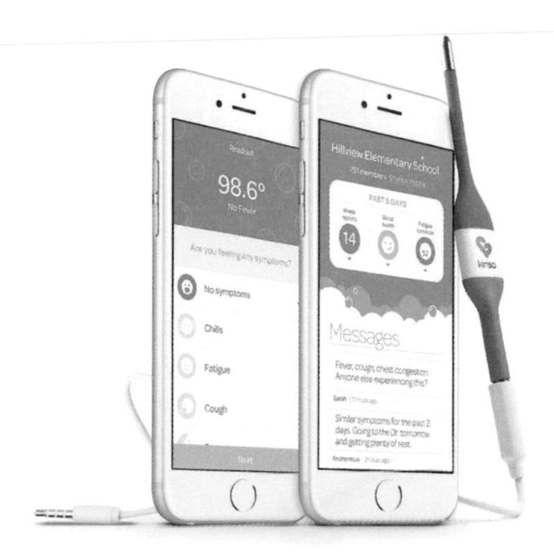

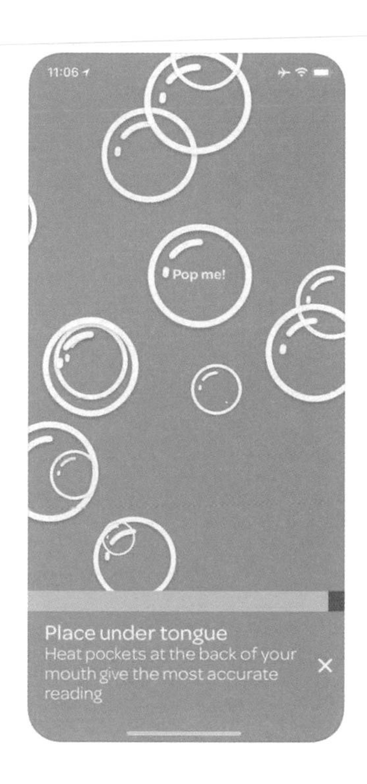

KINSA SMART THERMOMETER

Kinsa's excellent app was originally crowdfunded on Indiegogo and includes a simple bubble-popping game to entertain you while you wait for the result.

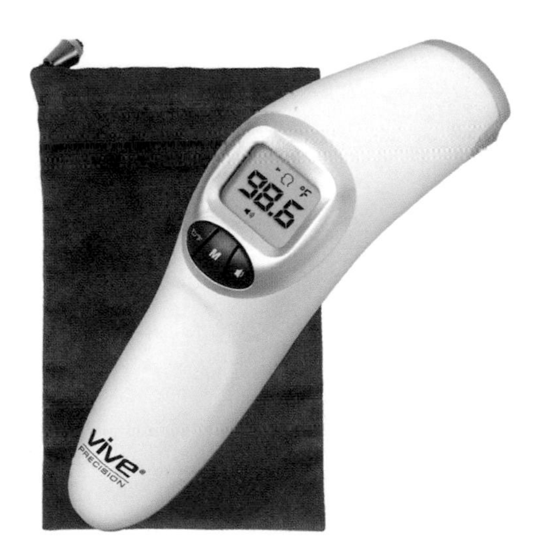

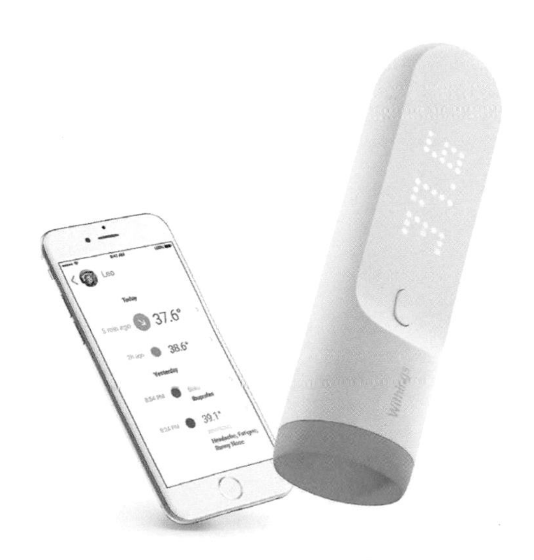

DIGITAL FOREHEAD THERMOMETER

Not '"smart," but it can give a reading in one second without even touching the subject.

TEMPORAL THERMOMETER

Nokia's acquisition of Withings added a smart thermometer to their product line.

BOTS & SEQUENCING

So far we've mostly concentrated on devices whose smart technology gives you the ability to switch things on and off differently, or use them in a new way, but beyond that there haven't been dramatic changes. However, the technology does exist to get devices working harder and being a lot more mobile.

FLOOR CLEANERS

Not as good, and more expensive

That's not just my assessment, that is the general view of the industry, which never really claims to be providing devices better than those they are replacing (but remains happy to offer machines that can cost considerably more than a laptop).

This level of honesty is actually quite refreshing; if you're choosing a robot vacuum, you might be doing it because you're less able to handle the cleaning on your own anymore, although many do so simply because it's an interesting piece of technology to have around — a mobile talking point that cleans the floor as it goes.

These robotic vacuum cleaners offer smaller capacities than uprights, and are pretty slow, though this problem should be offset by the fact you can get them to work while you're not around, and they can navigate beneath furniture that is usually a pain to get under, if not outright impossible. You'll still need to empty them, so you have to visit the device and its "home" fairly regularly. And that home is another issue; docking stations are essential for the robots to return to and charge themselves (which, by and large, they are capable of), but docking stations are not a beautiful addition to your home.

There are two ways you can think of your robot vacuum: one is as a replacement for the upright that you fetch out of the cupboard and set to work, returning both the dock and the machine back in the closet when you're done. The smarter, and pricier, devices, however, definitely offer you more if you can find somewhere to keep their charging stations. The Neato D7 even shares what it has learned about your home's layout with you via the app, so that you can zone off areas with your fingers. Other devices can use virtual walls — in the form of physical markers you can place in your room, like magnetic strips (used by Xiaomi, for example) that go under the carpet or doorframe.

Stairs pose something of a problem no matter how smart your vacuum is; there is no way you can remotely send one up them. In terms of avoiding harm, almost all can detect and avoid the edge at the top of a staircase (anti-drop sensors are built in just as collision detectors are), so it'll simply be a matter of moving the machine and base station upstairs and reassigning it.

NEATO ROBOTICS BOTVAC D7

A connected Wi-Fi robot vacuum cleaner with intelligent memory mapping.

IROBOT ROOMBA 980

A robot vacuum that can plot its path around obstacles like table legs, and can be scheduled with an app.

The worst thing about vacuuming isn't whizzing around with the machine – it's preparing the surfaces (especially if your kids have thrown toys everywhere). At best a robot will simply avoid them. Really, though, unless you keep your home like a catalog, the robot vacuum solves only the less-irksome part of cleaning the home.

MODE	EXPLANATION
Auto	The robot will set its own path and use sensors to avoid obstacles.
Manual	Steer the robot yourself using an app or a remote control.
Point clean	Send the robot to a particular area (such as a spill).
Turbo	Sacrifices battery life (and quietness) for power.
Dust sense	Cleans the room and spends longer in the areas where sensors detect dirt.

LAWNMOWERS

Also not as good, and also more expensive!

Just as with robot vacuums, there is an argument that devices in this category are not as good and are considerably more expensive than human home help. That, admittedly, is a shortsighted view; even if the prices don't fall, your grass isn't going to stop growing, and paying for a service adds up.

But I'll bet that wasn't the first thing you thought about. If you're anything like me, the first image that went through your head was the terrifyingly powerful sharp metal blade required to slice through blades of grass and the potential for error. A vacuum going astray is one thing, but a lawnmower? That's the stuff of horror movies.

Why would you empower the future army of the robots with something so risky as a lawnmower? Because, assuming Armageddon has not yet come and you're able to resist the temptation to put your arms underneath it, it also offers by far the gentlest way to mow your lawn. Rather than allowing each blade of grass to grow until you have time to cut them, then cutting off a lot every time, the robot mower takes a small clip more often, which is better for the grass.

Since robot lawnmowers tend to be electric and operate from a battery, they are usually quieter. The reduced power is not too much of a problem because there isn't the same imperative to be finished quickly as there is when it's just another of your chores, and the idea is that the robot runs more often, doing less work each time.

At the time of writing, all of the robot lawnmowers on the market are also mulching models, which means that they finely chop and scatter the cuttings over the lawn, thus returning nutrients to it, which in summer helps prevent excess evaporation. These factors are sure to be emphasized in the mower's promotional materials – what they're less likely to mention is that this also returns weeds and their seeds to the soil, while a mower with a collection bag removes them. Ultimately, you need to make the decision about whether you want this kind of clipping or not.

Installation of these devices is difficult, too. More so than a robotic vacuum cleaner, which will use its onboard sensors to detect obstacles. The lawnmowers also have sensors, but to ensure reliable edge detection, you'll need to lay a cable around the area that is to be mowed. The cable will take a few weeks to disappear under the grass. You can use a spade to make a thin crevice and bury the cable, but it's probably better to use pegs at first (and don't skimp on them – it's important that the cable isn't loose enough to rise up and get cut).

You might also think that the mowing method is a little bit, well, dumb. The Flymo 1200R (one of the more accessibly priced robot mowers) operates by driving forward until it senses the edge wire, then randomly redirects itself. Others do attempt to operate in lines, but achieving the traditional striped effect might be a bit much to ask.

FLYMO

The 1200R offers automatic mowing once you've set up a guide cable; the grass grows over it in a few weeks. When mowing there are plenty of safety sensors that will stop the machine quickly; small, light, razor-sized blades on a plastic central hub mean the machine can stop faster than a traditional mower should it sense a danger.

SENSOR STRIP

The route of the buried strip; it is father away near the lower flower bed than at a same-level path (because if it falls into the flower bed, it will get stuck). It also has to take a long way around the tree.

PET CARE

From food dispensers to robot balls

Looking after your pets has never been harder in an age when you can't be sure whether you'll have to work late, or simply fall at the mercy of the transport system. At the same time, you want the best for your pets, so smart homes have a number of options to consider.

In fact, the demand for this kind of product has been demonstrated by a number of successes on crowdfunding platforms like Kickstarter, suggesting enthusiasm for these products comes from a demographic below the age of 44 and is skewed slightly toward men; whether that's true of pet owners in general is another matter, but it's interesting to note. These tech-savvy pet-loving backers are responsible for offering everyone the opportunity to remain close to their furry friends, wherever they are.

We've already looked at cameras in some detail, in particular, cameras for the baby-monitoring market. Those who feel that pets are really baby substitutes might have their suspicions bolstered by seeing that very similar technology is offered, under different brands, to pet owners. Again, these products make use of the ability to remotely hear and speak to your pet through the camera. They've been adapted for the pet market with the added ability to dispense (dry) food from a compartment within the machine for routine feeding or to reward your friend.

Robotics can help with more than mere sustenance, too. The ball-shaped Pebby uses its outer shell as a wheel and contains a built-in camera so it can be remotely guided around your home using an app. It's as easy to operate as a very simple video game (just one directional control), or alternatively the ball can play with your pet itself, following a Wi-Fi device on its collar. The camera is obviously subject to a good bit of vibration when it is traveling at speed, but it's definitely effective with a playful dog or cat.

PETSAFE DOOR

With a simple chip added to you pet's collar, you can make your cat or dog flap unlock itself only for him or her.

SUREFLAP

A microchip-based cat flap, but this one can also be connected to a hub that relays your cat's movements to your phone app — to tell you whether they are in or out — and can be set to be locked or unlocked, or to lock itself after your cat comes home.

PETCUBE BITES

This is an HD camera not unlike some of those sold for security applications, but it adds a treat dispenser which can throw a pet treat from a built-in compartment.

Not that much bigger than a tennis ball, the worry would be how quickly a pet would get bored with Pebby, but at least you have the option of it interacting with the animal. Like a robot vacuum, it can also head home to a (very elegant) base station to charge (although you may be wondering just how many of these stations the modern apartment can sustain).

Petcube offer a; a monitoring camera with a remotely controllable laser pointer to distract your pet. Great fun until Mr Fluffles walks out of shot . . .

On which subject, the means of recovering a lost pet has also seen a significant number of gadgets emerge. Size and battery power are issues with these devices, which are typically attached to a collar – not to mention remembering to charge them. GPS has the advantage of global coverage, but the downside is that it only really works outdoors and requires power and a signal to return data (so likely a contract like a mobile phone). Alternatives are based on the Tile approach (not a pet-specific product), using long-life batteries — but these can only be tracked within the range of a Bluetooth signal. This is up to 200 feet (60 meters), with emphasis on the "up to," as any solid objects in the way will affect the range. A male cat's typical personal territory will be a 1500-feet (500-meter) radius from a center point (probably your home), so you may still need to wander a bit to find your pet.

PET TRACKER

Whistle's Pet Tracker not only helps you locate a missing pet, but also tracks activity in a similar manner to a smartwatch or Fitbit, so you can monitor your best friend's health.

TILE

The Tile lost things locator is an option for a pet. You need to put the tag on their collar, and it'll pin them down once you're within Bluetooth range. This is better for cats, which tend to wander shorter distances than dogs.

PEBBY

A ball with motors inside and a camera so you can play with your pet remotely.

TABCAT TRACKER

A variant on the Tile approach, but offering roughly double the range. Instead of using your phone, the credit-card-sized detector is simply waved and beeps.

ELECTRIQ BALL LAUNCHER

A toy that will "throw" a ball for your dog to fetch and — when the ball is dropped back into the top — can dispense a treat.

SOCIAL ROBOTS

When Alexa and her ilk get a body

The future for robotics in the home seems not quite to go all the way to a robot housemaid (or Terminator) just yet. Instead, consumer electronics shows have started to play host to diminutive in-home buddies that seem like slightly more alive versions of the more mainstream smart speakers.

The Jibo (pictured opposite) costs as much per month as Amazon's Echo Dot does to buy outright, and with a lot less functionality, it's really more for the kind of person who's more interested in playing a role at the vanguard of future tech than simply making their lights voice-activated. Jibo is able to interact with a number of the same devices as Alexa and friends, but with a much smaller take-up, there is hardly the commercial incentive for device developers to ensure compatibility.

Jibo's creator, Cynthia Breazeal, when interviewed in *Wired*, was not unaware of the first-generation nature of her baby, and the issues that go along with that, but did point out the potential good of a device with which families can interact in a more communal way than individually staring at screens.

On the subject of interacting with humans, Jibo might not be as early on the scene as all that; back in 1999, Sony first shared their robot dog Aibo with the world, and after something of a lull they've brought him back, albeit a good deal more sophisticated (and with a price tag to match).

Aibo is unlikely to be a major earner for the company that brought the PlayStation brand to the world, but finally you can buy a four-legged friend that, after your first paw shake, will get closer to you over time using deep learning and its connection to the servers that provide it. It's fair to say that Sony won't be stopping here.

The technology is amazing, and while Aibo is certainly not as nippy as a real animal, that may well be ideal for you. The simulated eyes, which blink, wink, and change pupil size, as well as the animal's gestures, are undoubtedly kawaii (Japanese for "adorable"). Plus, of course, you won't have to pick up after him. For my money, his mobility makes him engaging and enjoyable, while the fact that Jibo remains planted and only able to dance on the spot makes me feel a little bad for him.

While there may not yet have been as much commercial interest in robots as developers might have liked (ASUS's Zenbo, which we saw on page 11, launched in 2016 in Taiwan but is still waiting to be introduced to the rest of the world), this is almost certainly going to change. Amazon currently has a home robot in the works, and if this takes off, we'll undoubtedly be seeing more blurring of the lines between digital assistant and robot companion.

Jibo's eyes literally follow you around. I'm not at all sure how I feel about that, and his jokes definitely don't make up for it!

JIBO THE SOCIAL ROBOT

Jibo has two "eyes" which it uses for facial recognition, as well as an array of microphones. It can turn its "head" as it chats with you (or twerks with you), while the touchscreen in the middle of its face provides an alternative way to interact (though, in his own words, "high fives hurt my face").

AIBO THE ROBOT DOG

The latest iteration of Sony's robot dog.

SEQUENCING & INTEROPERABILITY

Making different technologies work together

The promise of smart technology is not just that each individual appliance or gadget behaves in a clever way, or can be operated from your phone's screen, but that all of them will busy themselves to achieve your desired goal from a single command.

In practice, you might find that – at least in comparison to the theoretical possibilities – this is a little disappointing. That stems from the fact that multiple standards are likely to exist in any home with multiple smart technologies, and not all of them will be designed to take others into account (depressingly few of them, in fact). This is an issue known as "interoperability," and something well worth checking up on before you buy a product if you have grand designs for elaborate sequences.

The exception to that rule, of course, comes in the form of the key platforms we met at the beginning: Alexa, Google Assistant, and, to some extent, Siri. With manufacturers of products falling over themselves to offer compatibility with these already established platforms, they (and perhaps even Microsoft's Cortana) are in a good position to offer the additional ability to perform a number of tasks for you at once, from a single command and via a single app.

Both Amazon/Alexa and Google Assistant call this function "Routines." To create a new routine for Alexa, you open the Alexa app on your phone and choose Routines from the main menu, where – assuming the devices

CHOOSE "ROUTINES" FROM THE MENU IN YOUR ALEXA APP AND PRESS "+" TO CREATE A NEW ONE.

CHOOSE THE "WHEN THIS HAPPENS" TYPE, THEN CREATE A PHRASE LIKE "ALEXA, MOVIE TIME" THAT WILL TRIGGER THE ROUTINE.

GO TO "ADD ACTION" THEN SELECT THE "SMART HOME" MENU

CHOOSE ACTIONS TO PERFORM, FOR EXAMPLE 1) DIM LIGHTS, 2) TURN ON TV

REMEMBER TO SAVE YOUR ROUTINE

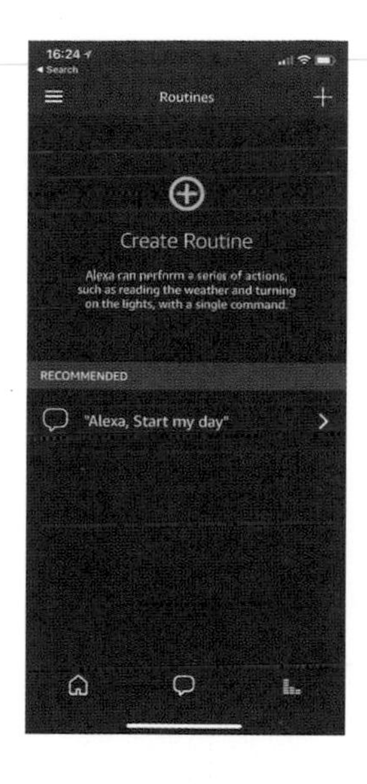

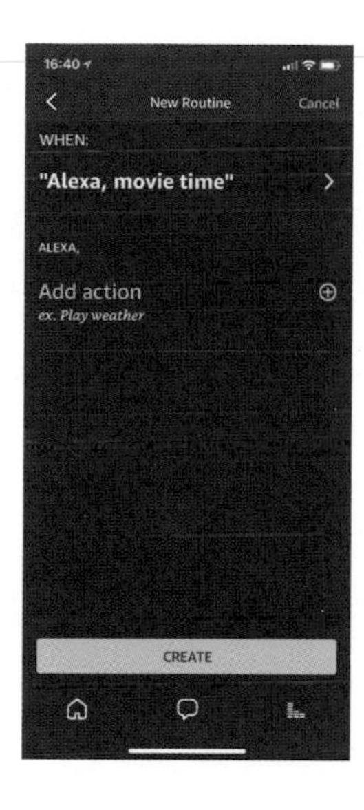

ROUTINES WITH ALEXA

This is how the interface appears when you create a Routine in the Alexa app.

in question are recognized by Alexa – you will find them awaiting addition to your sequence. A routine can also include Alexa functions that don't need a separate device, for example, it can read you the news and turn the lights on, and can be set to do so at a specific time rather than on a key phrase if you prefer.

Google Routines works in a similar way; you can either pick from preset options relating to specific times of day, e.g. "Good Morning" or "Commuting from work," with some likely actions suggested, or create a custom routine.

An alternative to Routines is Smart Home Groups (to borrow Alexa's term). These boil down to creating single names for more than one device so they can be operated at once. It's worth remembering that this is something your product-specific app might already have taken care of (all the lightbulbs in my living room are Hue bulbs, and the Hue app also has a Groups function; the groups I set up in Hue appear indistinguishably from a single Hue bulb in the Alexa app's list of devices). If you have mixed brands, this is a handy solution.

Siri takes an ever-so-slightly different approach, with "Shortcuts," which can be sequenced using an app that Apple added in iOS 12 (late 2018), and can operate some apps as well as HomeKit-compatible devices.

SEQUENCE WITHIN AN APP

A good example of a system with sequencing is Harmony, which can be configured to simulate pressing different buttons on different remote controls in any order you choose. It does this without using robot hands but its own infra-red emitter which you need to position in view of your devices and connect to your network.

IFTTT

If this, then that

When choosing between the many different smart-home devices, the inclusion a good app and/or compatibility with one or another of the voice assistants, are often the top priorities. But you will often find another platform listed among the "works with": IFTTT.

An acronym of "if this, then that," which is a conditional statement in programming, that's exactly what is on offer here; the opportunity to have one thing do something if another condition is met. In other words, you use it to create own bit of software with just one function. These mini-programs are known as "applets."

What's really interesting is that the triggers for your applets – the "if" – can come from a very wide number of things, not just smart-home devices. Similarly, the action that is triggeres — the "that" — might be anything from sending a tweet to turning on your lights.

Not that you need to write your own applets; once you sign up for the platform and commit to its privacy settings (it will need access to your smart home devices, among other things), you will also be able to pick from selections of ready-made applets that will be recommended to you based on your devices.

The only real downside is that it isn't as instant as you might hope; working as it does via the IFTTT central computers, things can take a few seconds — perhaps even minutes — to work. That, however, is just as much of an issue for competing platforms like Zapier. The latter also offers more possibilities (more than one thing can happen as a result of the process), but most people seem to find its interface harder to use.

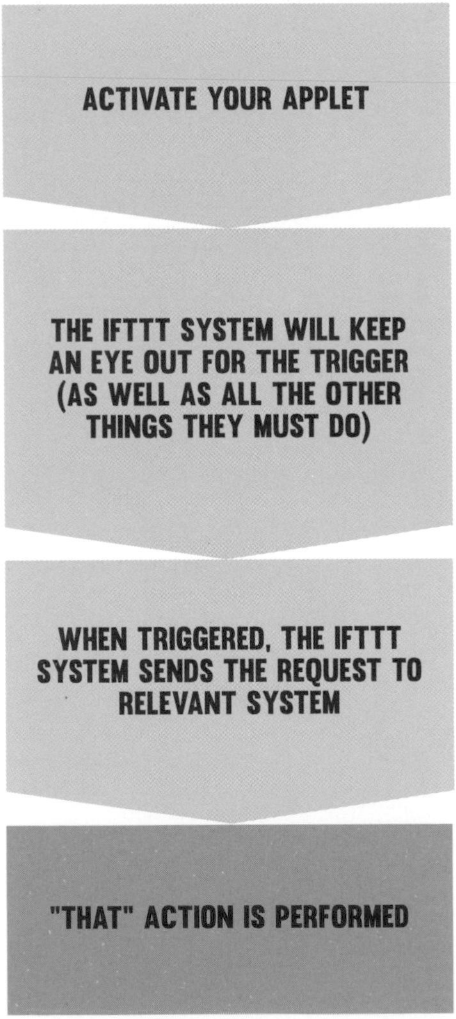

ACTIVATE YOUR APPLET

THE IFTTT SYSTEM WILL KEEP AN EYE OUT FOR THE TRIGGER (AS WELL AS ALL THE OTHER THINGS THEY MUST DO)

WHEN TRIGGERED, THE IFTTT SYSTEM SENDS THE REQUEST TO RELEVANT SYSTEM

"THAT" ACTION IS PERFORMED

I was a bit disappointed the first time I set one of these up; it wasn't difficult but I started with so many ideas and the speed of the system made most of them pointless.

IFTTT COLLECTIONS
You can browse for fun or useful applets created by others in IFTTT's easy-to-navigate interface.

DIY "ROBOTS"

You can use smart plugs to handle a robot's job

We already looked at the possibility of using smart plugs to turn lights on and off remotely, and as part of a morning routine. These clever little devices can be co-opted to do even more than that, adding up to a simple DIY 'bot.

Getting your morning coffee from a 'bot is undoubtedly less impressive if it's not delivered to you by a futuristic self-balancing robot arriving at your bedside like a more useful Segway, but it is also a lot easier to achieve, and the coffee will be just as good.

Ironically, this only works if you have a very modestly priced filter coffee system; anything more expensive will likely have a built-in clock, so all you'll find when you've rubbed the sleep from your eyes and walked to your machine is a blinking LCD display.

PUT A SMART SWITCH (E.G. WEMO) BETWEEN THE COFFEE POT AND THE OUTLET

USE ALEXA'S "FIND NEW DEVICE MODE" TO CONNECT TO THE WEMO

TURNING ON THE WEMO, VIA THE APP (OR ALEXA, ETC.), WILL TURN ON YOUR COFFEE

BEFORE BED, REPLACE WATER, FILTER, AND COFFEE, AND SET MACHINE TO "ON"

SMART PLUG

The WeMo Smart Plug is a good option for this kind of project because it doesn't need a hub and can work directly with your phone or an assistant like Alexa. It's also IFTTT-compatible for even more control.

COFFEE ON AUTO

It might not be the fanciest bit of technology you've ever seen, but smart plugs offer a relatively inexpensive way to smarten your home, and coffee addicts will love waking up to a freshly brewed coffee.

TRADITIONAL COFFEE POT

As long as the outlet power is on and the switch is set to "on," this machine will try to make coffee; perfect for DIY coffee bots (remember to add water and coffee!).

RELIABLE SETUP

Smart home technology means that your home's network, wireless and wired, is going to be handling a lot more traffic, and ensuring its healthy functioning is, therefore, what IT professionals like to call "mission critical."

CABLE MODEMS & ROUTERS

Should you take advantage of wireless systems?

Your Wi-Fi router, quite possibly built in to your cable modem, is the center of your network and your wireless network, one that all other devices in your home need in some way. Yet it's very often not a piece of technology chosen by you, but by your provider.

In fact, some providers even lease them, charging a monthly fee, which gives you a pretty solid incentive to invest in your own, unless you're planning on moving. (Cable modems conform to a standard called DOCSIS, but your internet may come via a DSL phone line depending on your area. Be sure you know how your internet is delivered and what box you may, or may not, replace.)

Now check what kind of connection speed you are getting. Your provider gives you one speed (typically the more you pay, the higher it is), but your router will also be speed-limited depending on what technological standard it uses. For wired networking there are essentially two speeds: Ethernet 10/100 (or "fast"), and Ethernet 1000 or Gigabit Ethernet ("faster" — with ten times the speed).

Wi-Fi standards are numbered 802.11, followed by a letter or pair of letters, though things don't follow the sequence you might imagine (see the table). With the "ac" standard fully established, you can easily see (from the "Time to move 16GB" column) why having the newest standard in your router can have a significant effect on moving files between devices in the home.

The ultimate bottleneck may be the connection to your service provider (traditionally labeled WAN, or Wide Area Network, on the back of the box), but if you share files between computers, or use a backup drive, they will be affected by the network speed without reference to your internet connection speed.

For that reason I strongly recommend ensuring that your router conforms, at a minimum, to Gigabit Ethernet, and has at least four ports (called RJ45), although smart home installations will almost inevitably require more ports. Gigabit (properly called "IEEE 802.3-2008") emerged in 1999 and was to all practical intents and purposes still the standard a decade later (it is backward compatible with 10/100), so be suspicious of wired products that don't offer this speed now.

The dedicated cables that Ethernet uses comprise eight thin cables twisted in pairs inside a single case, and as manufacturing standards have improved, and shielding from interference got better, it has made it possible for more data to be pushed down cables that look the same, with the same connectors. The current standards are Cat 5e or Cat 6. (Cat 3, incidentally, is dismissively known as "Voice Grade" and refers to old phone wires.)

If you've already got some smart home equipment, it's likely you've encountered a Cat 5 cable when connecting a hub to your router, for example, or at the very least the router to the DSL/cable modem. If it came

WIRELESS STANDARD	THEORETICAL MAX	TIME TO MOVE A 16GB FILE (APPROX 20 MINS OF 4K VIDEO)	FREQUENCY (GHZ)	ALSO COMPATIBLE WITH
802.11a	54 Mbps	39 mins	5	-
802.11b	11 Mbps	193 mins	2.4	-
802.11g	54 Mbps	39 mins	2.4	b
802.11n	600 Mbps	3 mins 30 secs	2.4 or 5	b/g
802.11ac	1.3 Gbps	1 m 40 secs	2.4 and 5.5	b/g/n
802.11ad	7 Gbps	20 secs	2.5 and 5	b/g/n/ac

USB

Not included with all routers, this port enables you to plug in a USB printer or hard drive and share it with multiple computers easily.

TO MODEM

Your internet service provider will likely provide a cable modem that outputs a signal using an Ethernet cable. Connect your Wi-Fi router to this.

TO DEVICES

These are the Gigabit Ethernet connectors that share your network via cables, for when you don't want to rely on Wi-Fi.

CABLE MODEMS & ROUTERS

from your internet provider, your examination of the device may have gone no further than looking at the sticker on the outside to discover the network name and password it came preset with. You shouldn't stop there, though, because there are many more things that can be tweaked if you spare a moment to examine your system's settings.

How can such a minimalist box allow you to adjust its settings? Through a computer's web browser, which makes it no more unfriendly to use than a website questionnaire (though admittedly one with a few unfamiliar phrases). You can reach this settings page by typing the router's IP address into the address line of your browser. If you don't know its IP, check your system's network settings and look for "Router" (on Apple devices) or "Default Gateway" (on Windows).

In the menus on the settings page, you should find access to features like guest networks (see the box opposite). You should also be able to see all the devices that are currently connected, change your password, and more.

If your device also serves as a router but you want to improve your home Wi-Fi network with a new system (like some of those seen here), you will also find a "modem mode" in the settings, which will enable you to use the device to simply pass the internet on. However, only do this if you'll have the ability to reconnect if you need to. Make sure you've got the necessary cables: lots of lovely thin new laptops need adapters as they lack an RJ45 (Ethernet) socket.

You can also check which wireless standard is being offered by your system. If your device is offering backwards compatibility with older standards, you could try disabling support for a, b, and g to free up frequency for devices with n- and ac-based Wi-Fi, which should speed everything up because none of the precious radio bandwidth is "wasted" having to help out older devices.

COMMON ROUTER ADDRESSES

If you don't know your router address, try typing one of these into your browser:

HTTP://192.168.0.1
HTTP://192.168.1.1
HTTP://192.168.2.1

GUEST NETWORKS

One feature on newer routers that can be useful for security reasons and for tidiness is the ability to create a "guest" network. This is an additional Wi-Fi network with a different name and password (or even no password, if you like) that you can offer to visitors. It doesn't have access to any of your devices, just the connection to the internet. Someone visiting will be able to log in on their phone, enjoy the speed, save their data charges and at no point browse through your hard drive, print to your printer, play with your lights, or do any of the other things that you might consider a bit too friendly.

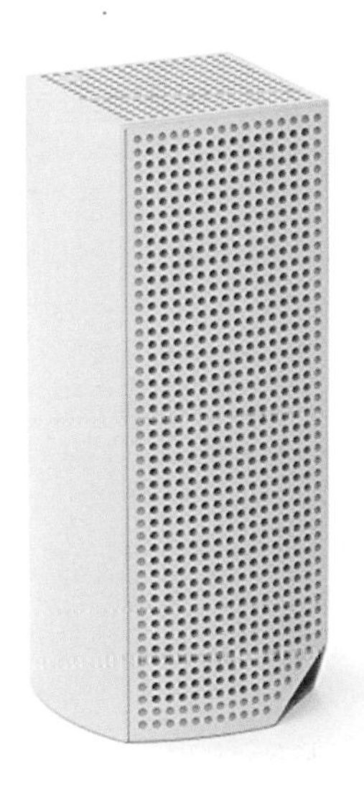

ALEXA-COMPATIBLE ROUTER

If guest modes and mesh networking weren't enough, this Linksys router, and some others, can be set up to respond to Alexa commands including "Tell Linksys to turn on my guest network." The password will be volunteered by the voice assistant too.

MESH WI-FI

These Orbi are not cheap, but for the money comes the ability to create a large Wi-Fi network with a single name. Unfortunately, the compact design allows for only one Ethernet socket.

TO WI-FI OR NOT TO WI-FI?

Should you take advantage of wireless systems?

The fact that more and more of the computers we use don't even feature sockets for network cables makes it seem like Wi-Fi is the inevitable future of connection, but to what extent should you depend upon it?

As we have seen, there are strong advantages to not using Wi-Fi technology, due to its range issues and high power consumption. A third issue is bandwidth; specifically, that there is a finite amount of it available. Older 802.11g has a limit of 20MB/s. If you connected two devices that both wanted 10MB/s at the same time you'd use that all up, but add a third and there is competition. The overall bandwidth will not then be shared evenly; Wi-Fi has a heavy range bias, meaning if one of those devices is a HD TV player right next to the router, it'll likely get a bigger share of the bandwidth than the laptop that is being used a couple of rooms away.

When high-quality video services are needed in rooms farther from the router, this problem becomes more obvious than it was, for example, when you were checking emails on a phone. Jerky, pixelated video is very likely to be caused by a poor connection.

You might think a dual-band router could at least partially solve this problem, but most are designed with a single radio chip that will handle one or another frequency at a time. Faster Wi-Fi obviously helps get the theoretical limits much higher (although the theoretical limit of every Wi-Fi standard has always been way above the practical results).

All along there has been a solution for moving data faster: swap Wi-Fi for cables. This also has the advantage of being considerably more secure. Some people are put off because cables can be untidy and fiddly, and those are legitimate concerns. But you can scale back your dependence on Wi-Fi without creating a whole tangled network of wires.

Powerline (sometimes called HomePlug) offers an elegant solution: plug it into an outlet and connect it to your router, and it will use your home's electricity cables to transmit a similar network signal to Wi-Fi. Your device connects via an Ethernet cable to another Powerline adapter at the other end and it will get a far faster data-transfer rate than if it were using your normal Wi-Fi network. Some models include a built-in Wi-Fi router, which you'd plug in at the far end as a means of extending your home's coverage. This is good if you want to go totally wire-free, but it might mean living with more than one network and password in the same house, which can be confusing and annoying for devices that you move between spaces.

The factors that affect the speed of Powerline systems are the quality of the wiring in the home and the system's own theoretical limit (some are advertised at up to 2,000 Mbps, but you should assume you'll get between 10 and 40 percent of that in practice). A clue to the truth of the speed might be the kind of Ethernet ports built in – if it doesn't say Gigabit, the Ethernet will throttle the speed to 100 Mbps.

CAT 5 ETHERNET CABLE

Eight cores of wire can move files fast.

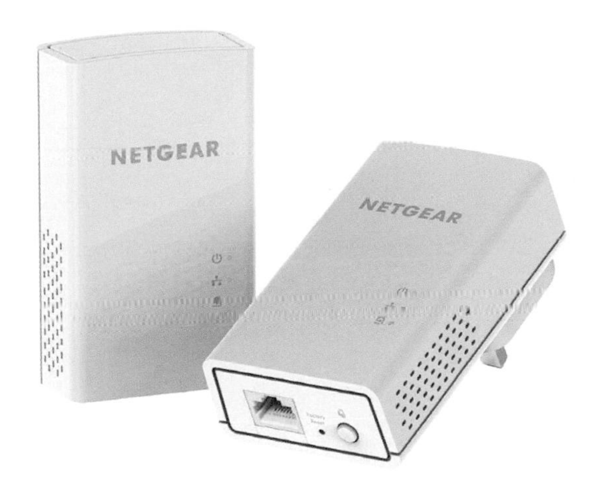

POWERLINE ETHERNET

Plug one in near the router and the other near the device you want to add to the network. Press the "pair" button at each end within a minute of plugging them in (you can pair them in nearby sockets then move one to make this easier), then use Ethernet cables to connect your device.

PASS-THROUGH

If you're low on outlets, look for a model with a pass-through feature like this one from TP-Link, which still allows you to use the outlet. (This is a UK outlet, but they are available for all standard sockets.)

BACKUP DEVICES

Protect your files with a shared drive

Managing files has become a great deal more complicated recently since "the Cloud" brought with it more options to consider. You can now store files remotely, have them backed up remotely from Dropbox, or, of course, keep them only on your computer or a physically connected hard drive.

Cloud-based services are useful, but often the space is expensive. While I'm writing this book, for example, I keep it in a folder provided by Dropbox for a monthly fee, because I can access the folder from any computer and use it in the same way. (As long as I remember to allow the computer enough time to synchronize the file over the network after I hit save . . .)

The more space I want to use like this, the more expensive it is, whoever the provider. But another issue is size; uploading and downloading large video files is not immediate and if you're working on a laptop in a cafe and suddenly need to leave, you don't want to have to wait ages worrying that closing the lid of the laptop mid-transfer will ruin the file.

To that end, backing up your computer locally, as well as simply saving files onto a disk drive on your home network, can be a much more palatable experience, and something a home network can make nice and straightforward — in terms of time at least. You need to make a one-off investment in a NAS (Network-Attached Storage) system, which is basically a hard drive with an Ethernet socket and a settings page you can access via your web browser.

The very best of these use more than one drive to back themselves up, a RAID (Redundant Array of Independent Disks) system. These can be configured to act either as a single giant drive, or in different arrangements so that if one of the individual hard disks that makes them up fails it can be replaced without any loss of data. Since there is a very established standard for hard drives, they are easy to source and relatively cheap to buy and replace. (Indeed, that is why RAID was originally developed, the "i" used to stand for "inexpensive.")

A word of warning — carefully read the guidance provided with your system if you're intending to use it with Apple or Microsoft's automated backup systems; they can be sticklers for settings. Personally, as someone who regularly stores video and high-resolution photos, my favorite thing about having a NAS is just having a whopping drive on which to store files, and I try to use an organized folder structure, creating a folder for each new year then each new month and each new day photographing. Some NAS systems have apps allowing you to view contents on an Android or an Apple TV; a feature worth looking out for, especially for video archivists.

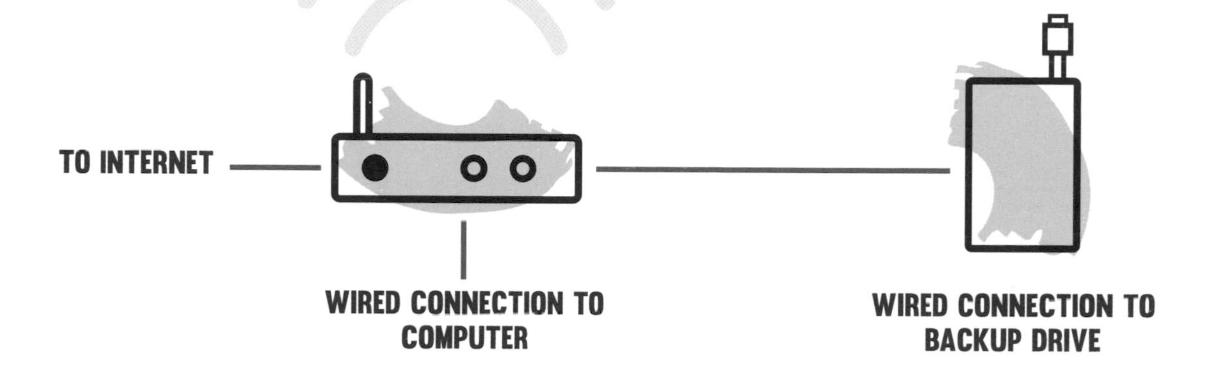

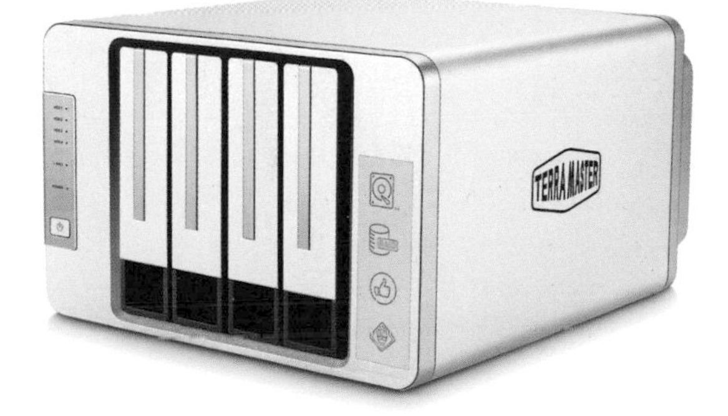

BACKUP BANDWIDTH

The faster your connections, the less of a chore backup and storage is, and the less likely you are to lose it. Connect your computer to the router with a cable connection, saving the Wi-Fi for the important backing-up work.

MULTI-DISK

A multi-drive NAS systems. Always check if the hard drives are included with the housing.

SECURITY & PROBLEM-SOLVING

Technology can go wrong, and the more smart home technology you have, the greater an issue that will be. However, you can avoid most of the potential problems by following a few simple precautions.

PLAN FOR FAILURE

Make sure you know what happens when the power is out

Traditionally, when the power was out, you would reach for the candles to light the home, knowing that the lights would just work again once the energy was restored. Now when the power goes, so do your networks, which can also mean your smart light switches.

To some extent, the designers of these smart home systems have your back. For example, if you normally turn on your Philips Hue bulb with one of Philips's networked switches, the bulb will still come on if you flick the traditional power switch a couple of times. You may lose the ability to dim or change moods until the mains-powered hub — and the network that it is connected to — has rebooted, but you won't find yourself in the dark. Similarly, if you only use the app (which also communicates with the bulbs via the hub), you will be left with the same level of control.

Some devices need their network connections more than others, and this is a big factor in which one you choose. It typically takes the network a few minutes to restore itself once power is reconnected to the essential devices (the router, the hubs, and so on). It is also possible you'll lose your internet connection, but in this case the internal networks will remain and you will just lose the abilities that connection affords.

A good example of the kind of device that is very dependent on both power and its internet connection is security cameras, which use remote servers for the intelligent examination and temporary storage of the footage that they capture. Without access to the systems, you will lose the notifications to your phone if a problem is identified, and since the cameras lack on-board data storage, you will also miss any incidents taking place during power or internet failures.

Voice assistants (Alexa, Siri, and so on) are also dependent on their connections to the internet, so if they get that through a powered network router that finds itself without power, they will be inaccessible until it returns. On the other hand, if you have access to these assistants on your phone or watch, they might automatically switch to a cellular connection. In that case you will be able to use the features that only require your phone and the internet (for example, setting alarms or getting the weather forecast), but not those that rely on your home's internet connection (telling the system to operate a device).

Other devices have been constructed with some level of built-in battery backing and the ability to conserve power. For example, Nest's thermostat and most of its competitors will stop using Wi-Fi when it is without power, so it can't be remotely adjusted, but it is able to continue controlling your heating and cooling systems (assuming power is reaching them).

Similarly, smart locks are virtually all based on battery power, so while the loss of mains power will make remote unlocking impossible, your key fob device should still operate normally. In that sense, smart locks should be viewed more as a handy bonus rather than as a real replacement; it's useful to have the ability to open your door from a distance, but it wouldn't be wise to entirely rely on that feature.

Finally, be mindful that security systems that use the internet to attempt to reach help require their connection, while the more traditional approach of automatically telephoning a security service or the police might continue to work, as telephone lines still connect without mains electricity.

BATTERY

At the moment a battery backup is still quite an extreme solution, but it is likely to become increasingly common as more people install solar panels. The dream of off-the-grid living is already perfectly possible in the US with the Tesla Powerwall, for example.

LAPTOP AND INTERNET

When your power goes, your router will likely go, too, unless you have some kind of backup power supply for your home. Since laptop computers and smart phones have built-in batteries, if you live in an area with good cellular coverage, you will still have all you need to use the internet for at least a couple of hours.

With that in mind, if you have not done so already, it's worth taking the time to set up "tethering." Tethering is the common term for using your cellular phone as a modem – usually with a wireless Bluetooth connection, despite the name. Once installed, you will find your phone available to connect to via your computer's Wi-Fi.

Your laptop likely uses a lot of data (web browsers often move a lot of data in the background, and programs automatically download files), so it is wise to have a good amount of data on your cellular account if you use this feature regularly.

iPhones and Android phones also use the term "personal hotspot" for this feature, and will guide you through setup.

HACKERS

The risks to your data and your systems

Hackers and hacking are much misunderstood, no doubt due to the portrayal in film and television of pasty goths sitting in basements looking at lines of code on screens, attempting to bring down governments.

Of course, it doesn't help that sci-fi-sounding words like "cybersecurity" are now correct terminology, with government departments so-named (with lots of names beginning with "prevention of," naturally). But just because all hackers don't conform to stereotypes about their appearance or goals doesn't mean that they don't represent a real threat. And by connecting devices to the internet, we do make it easier for them.

An internet-connected device isn't just at risk of being controlled remotely by someone you wouldn't want to take over, say, your light switches. Entertaining for some though that might be, there is little value in that for anyone. Some hackers do view their "work" as a public service, to demonstrate security flaws; one noted example in the smart home world was a group of hackers publicly streaming the video from baby monitors and talking to other people's children. In a way this kind of thing is useful, but it's harder to appreciate the service they're doing when it's your child who is being used as the example.

Another significant risk is that connected devices generally have a small amount of computing power of their own and could be persuaded to use that power for nefarious means. A common goal among hackers with a political agenda is a Distributed Denial of Service (DDoS) attack. It's not especially imaginative, but if you dislike an organization for one reason or another you might wish to damage its reputation in the eyes of others, or prevent it from doing business online. Since all websites or other connected computers do is respond to requests from other computers, all that is needed to stop them working is for too many requests to come in at once. Few people have the means to do this on their own, but if they can persuade lots of other computers elsewhere to make a request at the same time (to distribute it), they can achieve more harm.

Finding a way to send those requests via many IoT devices is one solution, and one reason why you need to protect your network. DDoS attacks that use your devices will slow down or damage your systems because they will use your internet connection.

Arguably, though, a more serious issue is someone attempting to gain access to your personal information so that they can then, for example, log in to your bank account and empty it. Or attempt to connect to one of your security cameras. It is even possible to remotely connect to a computer screen and see exactly what you are doing, which again risks divulging banking details if you're shopping online.

It's known that Samsung TVs (some of which have built-in microphones and cameras for video chat and voice commands) have been used as remote listening devices by hackers employed by the CIA, at least if you believe

WikiLeaks. (In a related note, courts have started to consider using recordings from smart home devices in evidence in serious criminal cases; it is not beyond the bounds of possibility that this begins trickling down to civil cases. In either situation, it also reveals that your personal conversations will be recorded on a server somewhere.)

HACK THREAT

Some hackers use their power to highlight security issues in technology; others use it just to cause harm. Even in the case of the former, would you really want to have your private data shared just to make a point?

That is not to say that the solution to all these threats is to stop using these systems. I was once a victim of a successful attack on my bank account but fortunately, having established I'd not done anything stupid like sharing the details on purpose, the bank refunded the money that was taken. It is definitely a good idea, though, to ensure that your system is secure and to be mindful that your system is only one of many where your information resides.

HOW TO STOP HACKERS

Take steps to prevent hackers using your system

For the most part, hackers will be deterred by some sensible precautions. Sadly, you can be sure that someone else won't have taken the same trouble, making them the easier target.

An important and simple first step is to ensure that your home Wi-Fi network has a suitably complicated password and, if possible, use a guest network system for visitors (see page 137). Although it will be a bit of a pain if you have already connected a number of devices, since you will have to log in and update the password settings, it is prudent to use a new password completely unlike the default one.

Hackers use a site, Shodan, to share the default settings of all kinds of devices, and it is easier to hack something if you know how many letters and numbers will be in the password, and if it starts with a letter or a number, for example. This principle is why so many websites ask you to include different kinds of character (letter, number, symbol).

That's not to say that something like "Pa55word" is OK either. On the one hand there is an uppercase "P" as well as lowercase characters, and two numbers, but it's also a pretty obvious solution. When a hacker is trying to break into a specific system they may use a computer program that tries one password after another. All such tools are written to try likely passwords (definitely including this, or the one you liked from *The X-Files*) first before starting sequentially.

Where your systems will allow it, using four entirely random words in a row, all lowercase, might actually be much harder for a computer system to guess than one word made up of an unnatural mix of numbers and cases (hat tipped to xkcd.com for this fact).

You should also go to your router settings and change the name of your Wi-Fi network (SSID) to something that definitely doesn't give away the brand of your router, your network provider, or your address. "Cybercrime Division B," or something, might do it.

Another router setting, as just mentioned, that can also make a difference is an additional guest network to keep devices and networks separate from each other. Finally, some routers have built-in Firewall software that can block internet access to all but needed ports and IP addresses. The downside of this is that you'll need to check which are needed by every new device you add, and those you already have.

One issue, though less likely to be a problem for those committed to smartening their home, is not completing setup processes. Remember that as soon as you have connected the device to your systems, it will start with the default settings, so even if you don't intend to use the smart features of a device straight away, if it is connected to your network it is actually better to follow its setup procedure, though.

UTM SYSTEM

A Unified Threat Management system is a pricey solution at the moment – one step up from a firewall – a small business might use a device like this to protect their entire internet connection from all kinds of hacks, unwanted content, and even blocking websites.

Hackers are creative, but every time they find out a new way to get into your system you can be fairly sure someone will be working to fix the issue (and, typically, the bigger the firm, the more confident you can be). Keeping your system's internal software ("firmware") up to date is important in this regard. Indeed, this is why you might sometimes get asked to install updates that seem to offer nothing new at all.

Don't neglect physical security precautions, either. Some people have been hacked because they reviewed devices on YouTube, and people watching their videos were able to make out their network settings and glean useful information from their footage. Others have the passwords written down in one place so that when even quite unsophisticated criminals break into their home they've been able to take the details away and empty bank accounts at their leisure. In some cases, routers with password stickers on have even been read through windows.

If you have children, it's also not a bad idea to think of them as being prospective "hackers." After all, they have pretty unique access to your systems! Make sure that one-touch purchasing and other features like this are disabled while you've got young children around; they don't even have to mean to spend your money to do so. From the age that they're old enough to press a button on a remote until at least the point they understand that money is a limited commodity, you should consider such features off-limits.

TROUBLESHOOTING

Some common smart home problems and solutions

When I ask Alexa (or another assistant) to turn on my lights, all the lights come on. I thought only the lights in the room that I was in would come on?
Initially it was necessary to specify the room to Alexa, even when you were physically in it. This is one of the reasons why it's an extremely good idea to name your smart home devices logically (see page 68). Since late 2017, it has been possible to go into Alexa's Settings menu, then Groups sub-section, and create a group associated with a specific Echo- or otherwise Alexa-enabled device. Add the relevant lights to the group (you can only have one group of each name) and from then on if you say "lights" to that device, it will assume you mean only the lights in the group.

If you are having this issue with Siri and HomePod, ensure the lights are in a group with the same name as the HomePod's location.

My IFTTT instructions don't seem to work.
Make sure that you checked how long the response from the servers can take. Many instructions take several minutes (and perhaps even more) between the "if" event and the "then" result taking place, so it might just be that you're not giving it long enough.

Someone who lives in my house is called Alexa. Can I still use Alexa?
Yes, but it's probably wise to change the "wake word" (see next query).

Alexa doesn't respond when I say "Alexa."
Assuming your device is plugged in and set up, perhaps the word Alexa isn't quite tripping off the tongue. Consider changing the wake word (the word that the device is listening for), which you can do by accessing the Settings page of the Alexa app, and call your device "Computer," "Amazon," or "Echo" instead.

Unfortunately, you can't change the way you activate Siri, and "OK Google" or "Hey Google" are the only phrases that work for Google so far, though it is rumored to be working on custom wake words.

I'm having trouble using the intercom feature on my smart home speaker.
Check that all of the devices are correctly set up; it is important that each understands there are others on the same network. On a Google device, try "OK Google, broadcast that we're leaving the house in ten minutes."

Alexa will do a similar thing with the "tell everyone" or "broadcast" keywords, but can also act as a direct room-to-room intercom. This only works if you have configured each Echo with the Drop In feature for both the home and the Echo devices in question, though, and if you've made sure they all have different names ("Kitchen," etc.). It also doesn't hurt to give them a software update then turn them all off and on. Unfortunately for Apple fans, Siri doesn't yet have this ability.

FIRST STEPS

If you're having problems with a specific device, you should always be able to turn to the manufacturer for support. Indeed, given the complexity of products available now, it is almost inevitable that you will need to do so. But given that we live in an age where much support is outsourced, sometimes to people who have never encountered the product, you can save time by trying a few steps first:

1. "Powercycle" the product. In other words, turn it off and on again, but do so properly with a shutdown option or power button, if there is one, and give it about 30 seconds.
2. If when it comes back on you are still having the same problem, check all the firmware is up to date. If not, update it. Check the same for any phone or tablet app you use to control it.
3. Powercycle the device and connected devices (if it is a hub, try to establish which of the things it connects to is causing the issue). It's also a good idea to restart the router, the hub of the whole network.
4. See if there are any possible network issues. Range extenders do sometimes cause problems. If possible, try the device from the original router.
5. Check the device name is correct, and doesn't conflict with another on the same network.
6. Check the battery of the device and all devices on the same mesh network.

TROUBLESHOOTING

I'm having trouble with a device's internet connection.
If a device is connected to the internet via Wi-Fi, make sure it's a good connection. Being too far away from the router can be an issue with Wi-Fi. One way you can get an idea about the quality of the Wi-Fi signal in different parts of your home is to use your phone. Download an app called Speedtest and first run it right next to the router. Then see if you get similar results in the location of your problem device (check that the phone is still connected to your Wi-Fi; you can check the IP address via the Results page of the app).

This technique is slightly unscientific; for one thing it is measuring the speed that is getting to your phone, not the device, and secondly, other devices on the network will draw some of that bandwidth. It says a lot about Wi-Fi that next to the router I was able to get 155 Mbps and at the opposite side of the next room (two walls away) the speed was just 27.7 Mbps. Speed isn't everything, but it does give an idea of the quality of the signal in the area. If it is poor, you can take steps to improve the range of your network. This might be as simple as repositioning your router (signal travels more easily through air than wall).

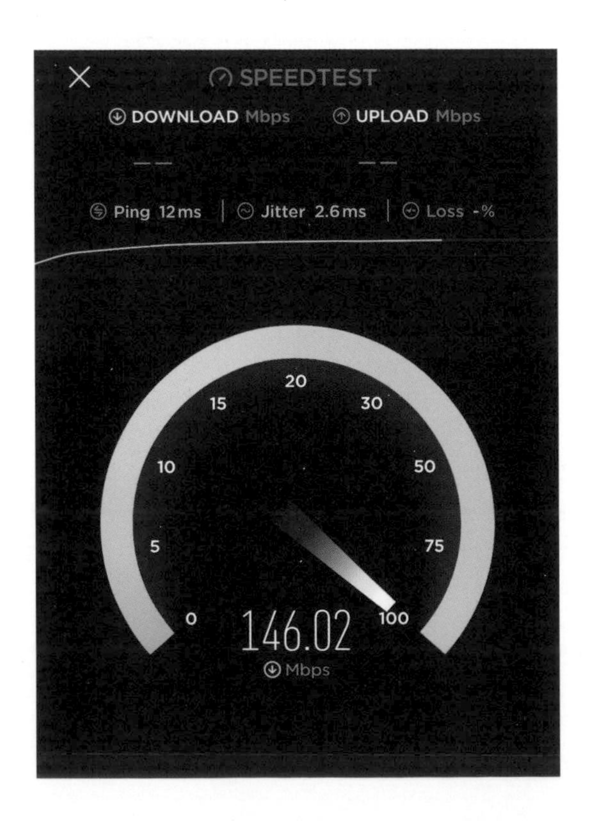

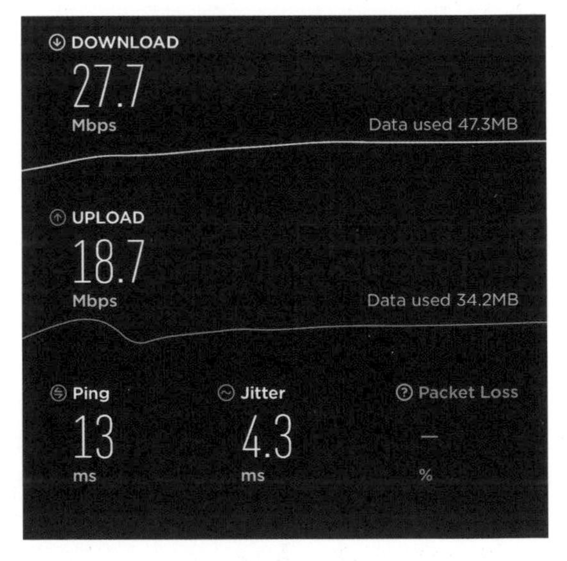

COMPARE & CONTRAST

On the left you can from see the animated measuring process that the speed is off the chart, tested at a point next to the router; the Results page, above, for the more distant test are less pleasing.

I'm having trouble with a device and it blames the internet connection, but this seems fine.
It is possible that the problem is at the other end; if the remote servers are down for any reason, or perhaps excessively busy, you might get the same error message on your device as you would when the connection failure was at your end.

I wish I could start again with one of my devices, as if it were new.
We've all got things up wrong before, and sometimes the idea of just starting again is appealing. Luckily, many products offer just such an option, possibly via a hidden button and/or a key combination, called a "hard reset." It's also your last resort before seeking manufacturer support, since it's often the first thing they suggest trying.

This reset might include all the firmware updates as well as the settings you have added, so be prepared for all your customization to be lost and plan accordingly. If you've discarded your original paperwork, Google "[name of device] hard reset" (include the model number, too) to find out exactly what will happen and how to do it properly. You'll likely also be able to track down a manual so you can follow the steps correctly.

When I ask my voice assistant to do something, it doesn't do it.
The most common reason for this is to do with the naming of devices. Use the app that comes with the relevant smart home device to make sure that it is set up using the correct name. If it is, make sure that it is actually compatible with your voice assistant without the need for a hub, or that you have the hub. Check the Settings page of your voice assistant's app to see if the system has detected the device on your network, and what the name is. An icon will represent the type (lightbulb, thermostat, etc.).

In the case of Alexa you might also find that you need to control it using a specific Skill — check whether you need to install a Skill, and if you need to mention it when addressing the assistant, e.g., "Alexa, tell Harmony to turn on the TV."

ARE YOU ALONE?

It's not for everyone, but there is one more thing you can try before giving up. Go to a search engine and describe the problem in a few words, being as specific as possible (use brand/model names). Your search engine may well now take you to a forum or community site where others have already addressed the issue, and either found a solution, or at least collectively drawn the conclusion that they're waiting on a software update from the manufacturer. This might mean scrolling through a few pages and researching what feels like a very confusing email chain, but it can sometimes yield results.

GLOSSARY

Some key terms and smart home jargon

Alexa
The default "wake word" and name of Amazon's voice assistant, but not the name of the device that features this functionality. Amazon's own-brand speakers are actually called Echo.

Bandwidth
A term generally used to describe the amount of throughput a data connection has, taking into consideration the amount of signal sent and the quality of the connection. Usually measured as "bits per second" (bps).

Bluetooth LE
A low-energy version of Bluetooth that is a common protocol for smart home devices.

DLNA: Digital Living Network Alliance
A DLNA is an agreed communication standard for devices established by Sony and Intel, now adopted by nearly all the major firms (except Apple, as you might have guessed).

Geofence
A virtual line set up so that when you enter it you trigger a function, for example, you can set one 650 feet/200 meters from your home that triggers the heating as you enter the zone, or another that turns off your lights as you leave it.

Group
In smart home settings it is often desirable to add numerous devices—for example, light bulbs—into single groups so that they can be switched on with a single command. Adding all the lights in a room into a group and calling that group Kevin means that you just need to say "turn on Kevin" and all the lights will be switched on at once.

Hub
A central device that allows different products to work together. This might mean different products in different categories, like Samsung's SmartThings, or a hub for one particular ecosystem, like Philips Hue, which connects little more than Philips's bulbs and Philips's switches, but it does connect them to your home's network connection so you can then operate them via an app on your phone.

HVAC
A corruption of Heating Ventilation and Air Conditioning and not actually a separate technology. A term more common in the US.

IoT: Internet of Things
A term used to describe the era of internet technology that is built into "things" to provide extra functionality, as opposed to merely in places we would expect to access it (computers, phones). As such the IoT and the smart home are very similar, though obviously the latter doesn't consider business or other applications.

IP: Internet Protocol
If someone says "that's an IP device," they mean that it communicates using standard networking (rather than, for example, a smart lightbulb that uses a smart home protocol).

Mesh network
A kind of network in which devices save energy and extend the range of the network by passing signals between each other (known as "hopping") rather than needing to communicate directly with a central point like a traditional Wi-Fi network. Z-Wave and Zigbee are mesh networks.

NFC: Near-Field Communications
This is the name for the technology that allows devices to communicate over a very short range — like contactless payment systems or the setup procedure for Apple's HomePod.

Outlet
If you're reading outside the US, you're probably more familiar with the term "socket."

Protocol
Wi-Fi, TCP/IP, Bluetooth, and numerous others are all protocols — agreed standards by which devices can communicate and software can be written to talk to them. Z-Wave and Zigbee are the first created with home-automation in mind since X10 in the '70s.

Remote access
Generally, remote in this context is operation from beyond your home.

Smart meter
Electricity and gas meters that share your usage with your utility company directly, and can also give you an idea of your own usage.

Voice assistant
Another term for digital assistant, but specifically referring to those designed to be spoken to, such as Siri or Alexa.

Wake word
The word or phrase which, when uttered, starts a voice-activated assistant such as Alexa or Siri ("Hey Siri"), or the Google voice assistant ("OK Google"). Typically they then send the rest of your command as a sound recording to be processed remotely by Amazon, Apple, or the assistant's developers, before being turned into a response.

Wi-Fi
The most common protocol for wireless communication in the home, allowing as it does fast-enough transfer ("bandwidth") for HD video (assuming a good connection). Wi-Fi is not suitable for all smart home devices, though, as it requires a relatively high amount of energy. That's fine for a device with a power supply, but if you want something fairly compact and discreetly sized, an alternative solution is needed (hence the mesh networks and hubs).

Z-Wave
A wireless protocol for the smart home that doesn't interfere with Wi-Fi or Bluetooth.

Zigbee
A wireless protocol for the smart home and other networks where long battery life is desirable.

INDEX

ACKNOWLEDGMENTS

"And I'd like to thank the Academy..."

Writing this book has been alternately a wonderful and a hugely frustrating experience, the former owing to my excitement of experiencing all the very best that technology can offer and the latter due to experiencing, well, everything else that tends to come with that. And then having to meet a deadline (cue coughing in background). Striking a balance between every changing range of "smart" products and my authorial desire to categorize them into settling trends has been enjoyable though more than a little taxing and I hope I did what you wanted. Strangely for me the most entertaining part was seeing all the power goes out very publically at the Consumer Electronics Show 2018; while I was there to research this book, that was the moment I knew for sure people really needed to see both sides of the tech promise.

The real heroes during my writing time, and to whom I dedicate the book, are Leandros and Vasi, my son and his mother, who have both endured an enormous additional workload at home while I hid in assorted coffee shops, drinking establishments and occasionally even the office while "completing" (we'll come back to that) this volume.

Writing a book is only part of creating one, and I owe a huge debt of gratitude to all my friends in ilex's orbit, including Roly Allen who helped me form the idea in the first place, Cassia Friello who came up with the design styling, Pete (another smart home enthusiast), Frank, Helen, Helena, Steph, and Ellie, but especially to Rachel Silverlight, the project editor (and unofficial commissioner) who actually bullied me into finishing the book and a series of rewrites which (now I've finished them) I'm happy to concede make for a better book.

. . . aaaaaaaaaaaaaaaand book!*

*This means more to Rachel, Frank, and me!

DO PLEASE VISIT

www.smartsmarthome.co

For any updates and product reviews, or to share your comments. It really is ".co" and not ".com."